Y0-AAG-209

GRAPHS AND APPLICATIONS

GRAPHS AND APPLICATIONS

Proceedings of the First Colorado Symposium on Graph Theory

FRANK HARARY
JOHN S. MAYBEE
Editors

A WILEY-INTERSCIENCE PUBLICATION
JOHN WILEY & SONS
New York • Chichester • Brisbane • Toronto • Singapore

Library of Congress Cataloging in Publication Data:

Colorado Symposium on Graph Theory (1st : 1982 :
University of Colorado in Boulder)
Graphs and applications.

"A Wiley-Interscience publication."
Includes index.
1. Graph theory—Congresses. I. Harary, Frank.
II. Maybee, John Stanley. III. Title.

QA166.C616 1982 511′.5 84-20819
ISBN 0-471-88772-2

Preface by F.H.

My gracious host in Boulder during the Fall Semester of 1982 was Professor John S. Maybee. His study of systems of linear differential equations led him to graph theory via the sign structure of the matrix of real coefficients. With his colleague Rich Lundgren of the University of Colorado at Denver, he has been applying graphs and signed graphs to economics and ecology.

My duties in Boulder consisted only of organizing the Ulam Chair Seminar in Mathematics on any topic of my choice. This topic was Graphs and Applications. The applications of graph theory which were presented included anthropology, chemistry, computational complexity, ecology, economics, operational research, and very large scale integration. These are somewhat reflected in this book.

After my departure from Boulder at the end of 1982, all of the workload of persuading this book to materialize fell upon Professor Maybee. With his staff he has succeeded heroically. I share the expression of gratitude in his Preface, which follows.

Frank Harary

Ann Arbor, Michigan
June 1984

Preface by J.S.M.

The First Colorado Symposium on Graph Theory was held at the University of Colorado in Boulder on October 15 and 16, 1982. It was organized and directed by Professor Frank Harary, who was then visiting the Mathematics Department where he held the Ulam Chair during the Fall Semester of 1982. This book includes the papers presented at the Symposium.

The Symposium, which was attended by 40 graph theorists, was spread over two days, allowing ample time to discuss the papers. The participants were also able to gather for two evenings at the residences of the editors, for further stimulating and fruitful interchanges.

At the present time, graph theory and its applications must be regarded as one of the most active areas of mathematics. This fact is attested to by the variety of topics covered here. Thus the editors are pleased to bring to the attention of all graph theorists, as well as to those who have occasion to use graph theory in their research, this collection of papers.

We would like to take this occasion to thank those who were involved in this project. First we are grateful to all of the authors, who were very cooperative in submiting their manuscripts promptly and who also responded to the need to quickly and accurately review camera-ready copy. An enormous quantity of gratitude goes to Mrs. Marie Kindgren, who typed all of the copy. She also kept detailed records of the status of each paper, and this was very helpful to the sometimes careless editors. Thanks go to Carol Maybee, who planned, organized and served a cocktail buffet on October 15 so that all of those who attended the Symposium could continue their informal discussions. Finally, we wish to thank the people at John Wiley & Sons who helped in the production of this volume. First there is David Kaplan, Editor, and his assistant, Lisa Culhane. Thanks go also to Rose Ann Campise, who presided over the production process.

John S. Maybee

Boulder, Colorado
May 1984

Contributors

Jin Akiyama

Department of Mathematics
Nippon Ika University
2-297-2 Kosugi, Makahara-ku
Kawasaki 211, JAPAN

Marlow Anderson

Department of Mathematics
Colorado College
Colorado Springs, CO 80903

Lowell W. Beineke

Department of Mathematics
Indiana University-Purdue University at Fort Wayne
Fort Wayne, IN 46805

Frank Boesch

Department of Electrical Engineering
Stevens Institute
Hoboken, NJ 07030

Gary Chartrand

Department of Mathematics
Western Michigan University
Kalamazoo, MI 49008

Hiroshi Era

Department of Mathematical Science
Tokai University
Hiratsuka, JAPAN

R. J. Faudree

Department of Mathematics
Memphis State University
Memphis, TN 38152

Frank Harary

Department of Mathematics
University of Michigan
Ann Arbor, MI 48109

Joan P. Hutchinson

Department of Mathematics
Smith College
Northampton, MA 01063

Brad Jackson

Department of Mathematics
University of California
Santa Cruz, CA 95064

K. F. Jones

Department of Mathematics
University of Colorado at Denver
Denver, CO 80202

Mikio Kano

Akashi Technological College
Akashi, 674
JAPAN

Marc Lipman

Department of Mathematics
Indiana University-Purdue University
at Fort Wayne
Fort Wayne, IN 46805

J. Richard Lundgren

Department of Mathematics
University of Colorado at Denver
Denver, CO 80202

Bennet Manvel

Department of Mathematics
Colorado State University
Fort Collins, CO 80523

John S. Maybee

Department of Mathematics
University of Colorado
Boulder, CO 80309

F. R. McMorris

Department of Mathematics
Bowling Green State University
Bowling Green, OH 43404

John Mitchem

Department of Mathematics
San Jose University
San Jose, CA 95192

Michael Plantholt

Department of Mathematics
Illinois State University
Normal, IL 61761

Gerhard Ringel

Department of Mathematics
University of California
Santa Cruz, CA 95064

C. C. Rousseau

Department of Mathematics
Memphis State University
Memphis, TN 38152

R. H. Schelp

Department of Mathematics
Memphis State University
Memphis, TN 38152

Edward Schmeichel

Department of Mathematics
San Jose State University
San Jose, CA 95192

Allen J. Schwenk

Department of Mathematics
U. S. Naval Academy
Annapolis, MD 21402

G. J. Simmons

Department of Mathematics
Sandia National Laboratories
Albuquerque, NM 87185

Ralph Tindell

Department of Mathematics
Stevens Institute of Technology
Hoboken, NJ 07030

F. C. Tinsley

Department of Mathematics
Colorado College
Colorado Springs, CO 80903

Stanley Wagon

Department of Mathematics
Smith College
Northampton, MA 01063

John J. Watkins

Department of Mathematics
Colorado College
Colorado Springs, CO 80903

U. M. Webb

Department of Mathematics
University College
Cardiff, Wales, UK

Robin J. Wilson

Mathematical Institute
Oxford OX1 3LB
ENGLAND

Contents

PATH FACTORS OF A GRAPH

Jin Akiyama

Tokai University
Hiratsuka, 259-12, Japan

Mikio Kano

Akashi Technological College
Akashi, 674, Japan

ABSTRACT. Our purpose is to propose a new viewpoint for graph factors, apart from the traditional degree conditions. A spanning subgraph F is called a path-factor if each component of F is a path of order at least two. In particular, a path-factor F is called a (P_2,P_3)-factor if each component of F is either P_2 or P_3. A P_n-factor F, for some fixed $n \geq 2$, is a factor such that every component of F is a path P_n of order n. Several results on (P_2,P_3)-factors, P_3-factors and P_4-factors and their applications for the "triominos tiling problem" are presented and also some graph decomposition problems related to these factors are discussed.

1. INTRODUCTION

We deal with only finite, simple graphs, which have neither multiple edges nor loops. All notation and definitions not given here can be found in Harary [13].

Let $G_1, G_2, \ldots, G_n$ be nontrival graphs. A graph G has a $(G_1, G_2, \ldots, G_n)$-subgraph H if H is a subgraph of G and each component of H is isomorphic to one of the G_i $(i = 1,2,\ldots,n)$. In particular, such a spanning subgraph of G is called a $(G_1, G_2, \ldots, G_n)$-factor of G.

A factor F is called a path-factor (or a cycle-factor) if every component of F is a path (or a cycle). From this point of view, ordinary 1-factors (or 2-factors) are just the same as P_2-factors (or cycle-factors). Those factors defined in this manner are called component factors of a graph. Several component-factors concerning path or cycle are listed in the followings.

A List of Component-Factors related to Paths, Cycles or Stars

1. P_2-factor; 1-factor, see Tutte [16].
2. P_3-factor
3. In general, P_i-factor for some fixed $i \geqq 2$.
4. (P_2, P_3)-factor; (1,2)-factor, see Akiyama, Avis and Era [3] or Akiyama [1,2].
5. $(P_2, C_n | n \geqq 3)$-factor, see Tutte [17], Hajnal [12] and Berge [9, Theorem 3.1].
6. $(P_2, C_{2n+1} | n > 1)$-factor, see Cornuejols and Pullyblank [10].
7. (P_2, C_3)-factor
8. cycle-factor; 2-factor, see Belck [7].
9. star-factor, see Amahashi and Kano [6].

2. (P_2, P_3)-FACTORS

A spanning subgraph F of G is called a (1,2)-factor if each vertex of F has degree 1 or 2 in F. A criterion for a graph to have a (1,2)-factor was discovered by Akiyama, Avis and Era [3]. The following three statements are equivalent.

(a) G has a (1,2)-factor

(b) G has a path-factor

(c) G has a (P_2, P_3)-factor

Hence, the criterion for (a) is the same as for (b) or (c):

Theorem A. A graph G has a (P_2, P_3)-factor if and only if

$$i(G - S) \leqq 2|S|$$

for every subset $S \subset V(G)$, where $i(G - S)$ denotes the number of isolated vertices of $G - S$.

This theorem has some interesting corollaries as follows.

Corollary A1. Let G be a graph with maximum degree Δ and minimum degree δ. If $\Delta/\delta \leqq 2$, then G has a (P_2, P_3)-factor.

Proof. Let S be a subset of V(G). Then the inequality $i(G - S) \cdot \delta \leqq \Delta \cdot |S|$ holds, and thus $i(G - S) \leqq 2|S|$ since $\Delta/\delta \leqq 2$. ▯

The next corollary follows at once from the previous result since $\Delta = \delta$ for regular graphs.

Corollary A2. Every regular graph has a (P_2,P_3)-factor. ▯

Every maximal planar graph has a (P_2,P_3)-factor although its ratio Δ/δ may become large.

Corollary A3. Every maximal planar graph has a (P_2,P_3)-factor.

Proof. Let G be a maximal planar graph, S be any vertex subset of G, and $I(G-S)$ be the set of isolated vertices of $G-S$. Then the neighborhood $N(G-S)$ of $I(G-S)$ is contained in S and the subgraph H induced by $N(G-S)$ of G forms a planar graph without end vertices, since G is maximal planar.

For any component C of H, we denote by $r(C)$ the number of regions of C. Then applying the Euler Polyhedron Formula, we obtain

$$r(C) \leqq 2|V(C)| - 4 \leqq 2|V(C)|.$$

Hence we have the following inequalities:

$$i(G-S) = |I(G-S)| \leqq \sum_{C\subset H} r(C)$$

$$\leqq \sum_{C} 2|V(C)| \leqq 2|V(H)| \leqq 2|S|.$$

Consequently, G has a (P_2,P_3)-factor by Theorem A. ▯

We introduce a family of graphs called triangle graphs. Let π be a plane with rectangular coordinates and L be a set of lines on π given by:

$$L = \{y = n \quad \text{or} \quad y = \pm\sqrt{3}\,x + 2n \quad | n \in Z\}.$$

An infinite graph I is obtained by taking the set of

lattice points of π as the vertex set V(I) and the set of unit segments of π as the edge set E(I).

A triangle graph is a subgraph of I obtained from a set of unit triangles on π, see Figure 1.

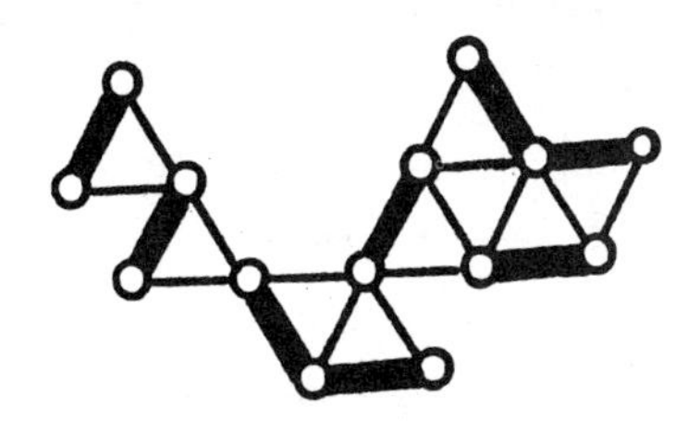

Figure 1. A triangle graph T and its (P_2, P_3)-factor

Corollary A5. Every triangle graph T has a (P_2, P_3)-factor.

Proof. Let S be a vertex subset of T. It is easy to see that for every vertex $x \in S$, the number of edges joining x and the isolated vertices of $G - S$ does not exceed 4. Hence we have the following inequalities.

$$2i(G - S) \leqq 4|S|. \quad \Box$$

3. P_4-FACTORS

We present the following theorem which is analogous to the theorem of Petersen which shows the existence of a P_2- factor for cubic bridgeless graphs.

The next lemma is required to provem Theorem 3.1.

Lemma B. (Berge [8, Theorems 6 and 7 in Chapter 18] and Plesnik [14]). Let G be an r-regular, (r-1)-edge connected multigraph of even order. Then there exists a P_2-factor containing an arbitrarily given edge e. ▯

Theorem 3.1. Let G be a 3-edge connected, cubic graph of order 4p. Then for any two edges e_1 and e_2, there exists a P_4-factor containing both of them.

Proof. A 3-edge connected, cubic graph G has a P_2-factor F_1 containing e_1 by Lemma B. Denote by G* the graph obtained from G by contracting every edge of F_1. (See Figure 2.)

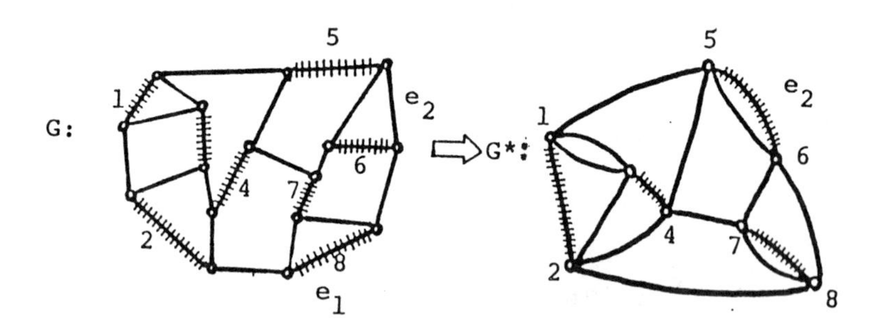

: a P_2-factor F_1 : a P_2-factor F_2

Figure 2.

Then G^* is a 4-regular, 3-edge connected multigraph. Applying Lemma B, G* has a P_2-factor F_2, which contains e_2 if F_1 does not contain e_1.

Then $F_1 \cup F_2$ constitutes a P_4-factor of G and it contains both e_1 and e_2. ▯

Corollary 3.2. Every 3-connected cubic graph of order 4p has a P_4-factor.

The graph illustrated in Figure 3 shows that the

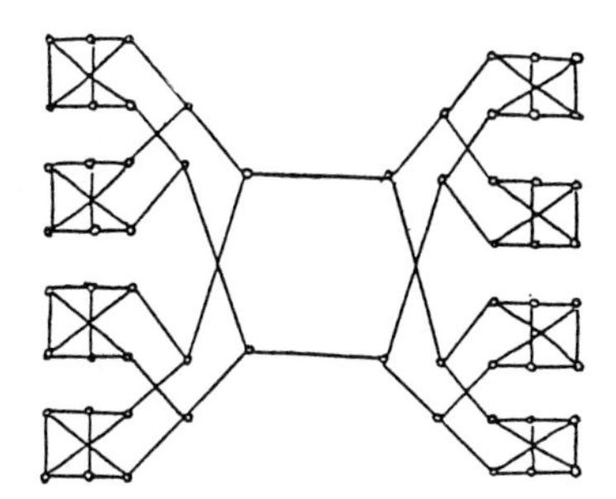

Figure 3. A 2-edge connected cubic graph of order 60 having no P_4-factor

connectivity hypothesis of Theorem 3.2 cannot be omitted.

4. P_3-FACTORS

We first discuss an easy necessary condition for a graph to have a P_3-factor.

Suppose G has a P_3-factor. Then for any vertex subset S of G, the components of $G - S$ can be classified into three types T_i $(i=0,1,2)$ according as their order i (mod 3). Denote by $\omega_i(G - S)$ the number of components of the type T_i, then we have the inequality (4.1) by estimating the least number of vertices of S needed to form a P_3-factor of G.

$$\omega_1(G - S) + 2\omega_2(G - S) \leqq 2|S| \tag{4.1}$$

However the condition (4.1) is not sufficient for a graph to have a P_3-factor, which can be seen in Figure 4(a).

We here propose a conjecture on the existence of a P_3-factor for cubic graphs.

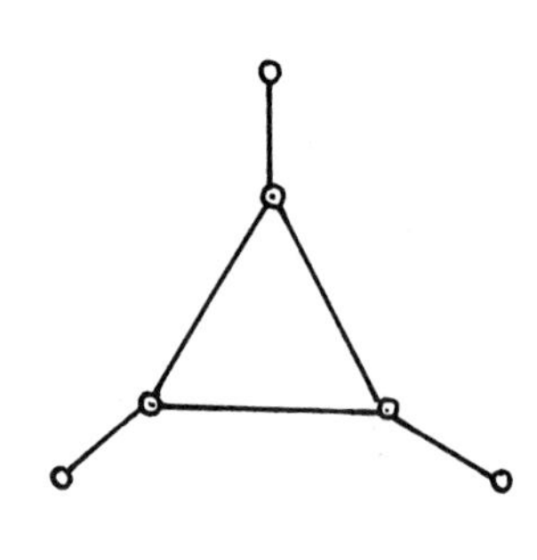

(a) A graph without P_3-factors, which satisfies (4.1)

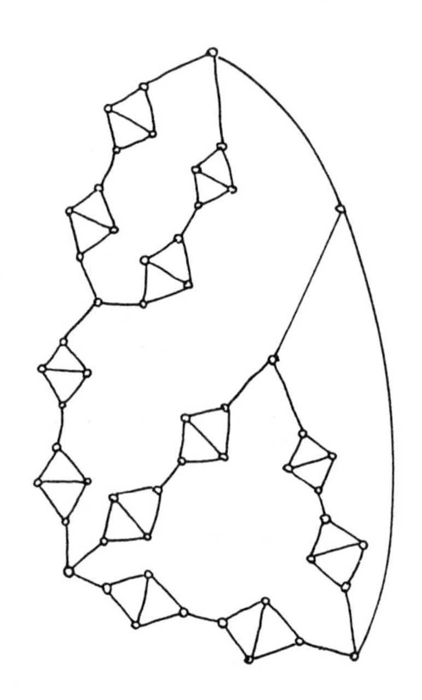

(b) A 2-connected cubic graph of order 54 with no P_3-factors

Figure 4.

Conjecture. Every 3-connected cubic graph of order $3p$ has a P_3-factor.

Note that there exists a 2-connected cubic graph of order $3p$ with no P_3-factors as illustrated in Figure 4b.

We shall introduce a special family of graphs, which is motivated from "the triominos tiling problem" in the next section.

Let π be a plane with rectangular coordinates and I be a set of the lattice points on π, that is, $I = \{(i,j) \mid \text{both } i \text{ and } j \text{ are integers}\}$. We define a graph G on the plane π as follows:

Take a finite subset $V \subset I$ as the vertex set of G and join every two vertices x, y of V if and only if the distance between x and y is 1. If a graph G obtained in this manner satisfies the property that G is connected and every edge of G is contained in some 4-cycle C_4 of G (which will be referred to as a square hereafter), then G is called a square graph, see Figure 5a. Note that the graph in Figure 5b is not a square graph.

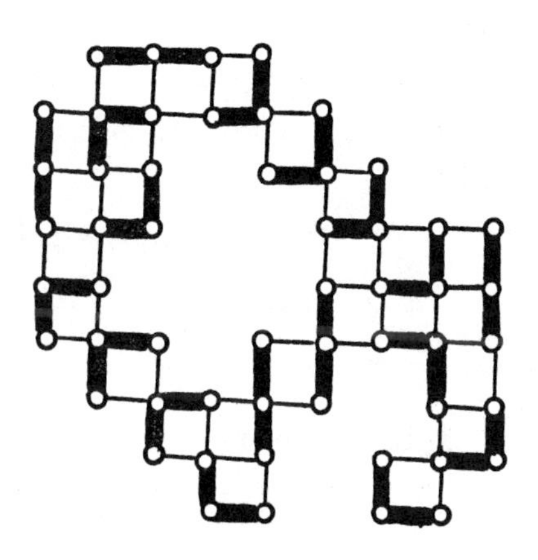

(a) A square graph and its P_3-factor

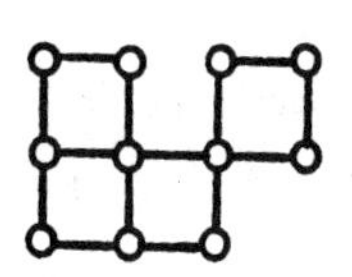

(b) A graph which is not a square graph

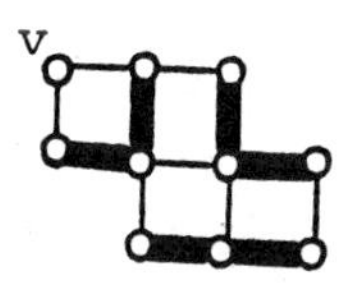

(c) G - v has a P_3-factor

Figure 5.

Theorem 4.1. Every square graph of order 3p has a P_3-factor.

Theorem 4.1 is an immediate consequence of the next theorem, for which we shall give a brief outlint of the proof.

Theorem 4.2. Let G be a square graph of order p.

(I) If $p \equiv 0 \pmod 3$, G has a P_3-factor,(Fig. 5a)

(II) If $p \equiv 1 \pmod 3$ and v is an arbitrary vertex of degree 2 in G, G - v has a P_3-factor, (Fig. 5c)

Outline of proof. For convenience, we say a v-semifactor of G is a P_3-subgraph which contains all the vertices of G except a single vertex v. By the capital letters A, B, C,..., we denote squares (faces) on the plane, and by the small letters a,b,c,..., we denote vertices of G. Two squares (faces) A and B intersect if they have a common vertex of G, and they are adjacent if they have a common edge of G.

Our proof is by induction on the order p of G, and it is shown in Figure 6 that all square graphs of small order p with $p \equiv 0$ or 1 (mod 3) have a P_3-factor or a v-semifactor respectively, as the base of induction.

(I) *G is a square graph of order $p \equiv 0 \pmod 3$.*

Lemma 4.1. If there is a square of G which is not adjacent to any other square, then G has a P_3-factor.

Proof. For convenience, we name the squares (faces) and vertices of G as in Figure 7a.

Suppose that x is not a cutvertex of G. Then $G-\{a,b\}$ has a c-semifactor by the induction hypothesis and so G has a P_3-factor.

Suppose that x is a cutvertex of G. If $G \not\supset C$, then $G-\{a,b,c\}$ has a P_3-factor by the hypothesis, and thus G has a P_3-factor. Hence we may assume that $G \supset B,C$. Set $G-\{a,b\} = H \cup K$ such that $H \supset C$, $K \supset B$. Then both H and K are square graphs. We now divide our proof into three cases.

Case 1. $|V(H)| \equiv 0$ (or 1) (mod 3) By the hypothesis, H (or K) has a P_3-factor and K (or H) has a x-semifactor (or c-semifactor), respectively. Therefore, G has a P_3-factor.

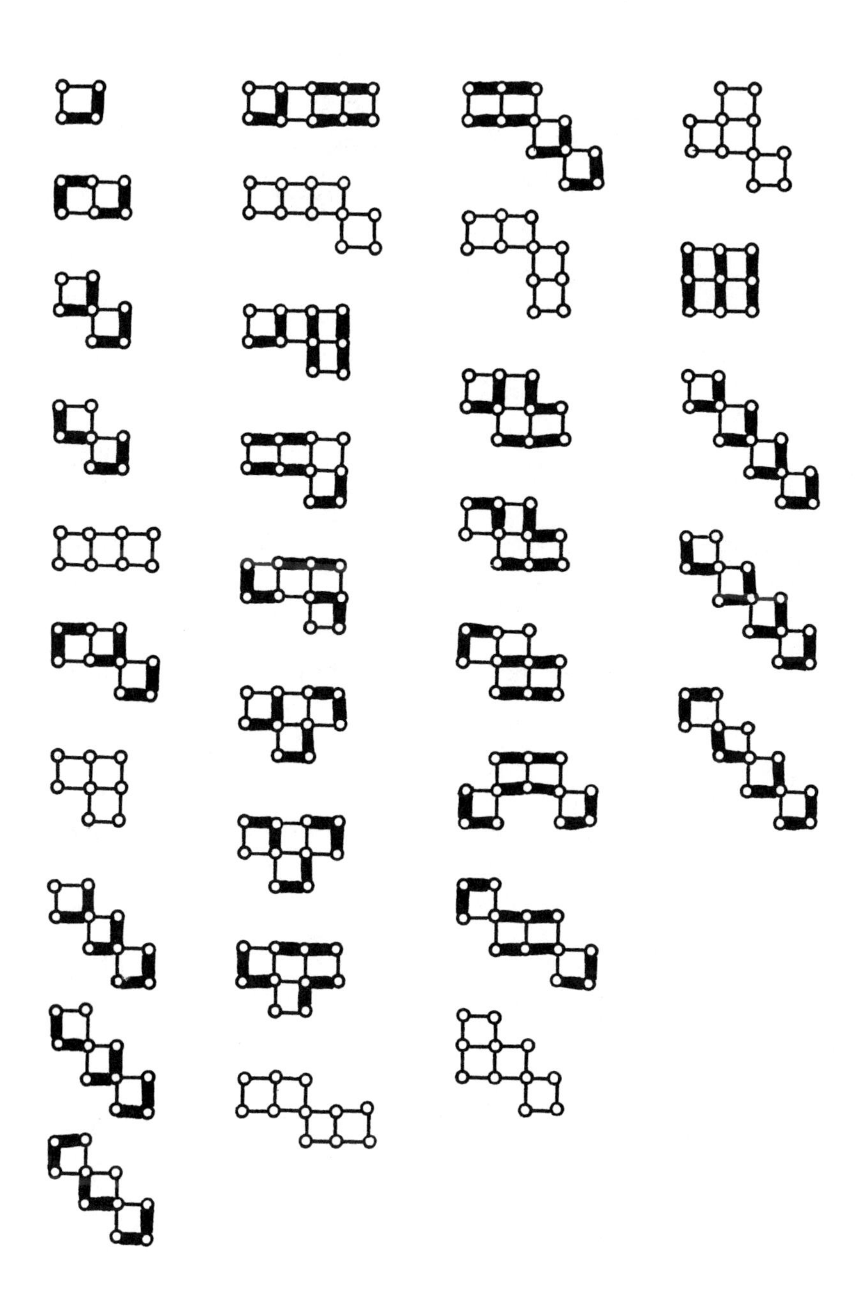

Figure 6. Small square graphs and their P_3-factors or v-semifactor when $p \equiv 0$ or 1, respectively

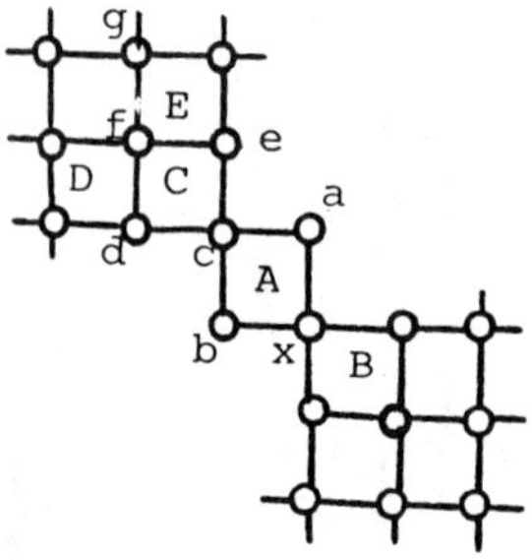

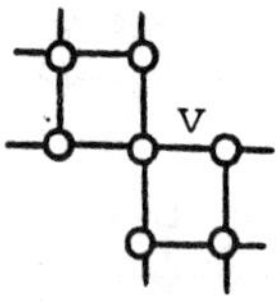

(a)

(b)

Figure 7.

Case 2. $|V(H)| \equiv 2 \pmod 3$ If $G \supset D$ and $G \not\supset E$, then $H-\{c,e\}$ has a P_3-factor. If $G \supset D$ and E, then $H - c$ has e-semifactor. If $G \not\supset D, E$, then $G \supset F$. In this case, it can be shown inductively that $H-\{c,d,e,f,g\}$ has a P_3-factor by inspection. Hence $H + a$ has a P_3-factor. Similarly, we see that $K + b$ has a P_3-factor. Consequently, G has a P_3-factor. ▯

We require three more lemmas in order to prove (I) of Theorem 4.2, but as their proofs are long and monotonous, we shall omit them.

Lemma 4.2. If G has a vertex v which is contained in exactly two squares (see Figure 7b), then G has a P_3-factor. ▯

Lemma 4.3. If G has a square which has exactly one adjacent square, then G has a P_3-factor. ▯

Lemma 4.4. If every square of G has at least two adjacent squares, then G has a P_3-factor. ▯

(II) G is a connected square graph of order 1 (mod 3) and v is an arbitrary vertex of G with degree 2.

Orientation of Proof of (II). For convenience, we name the square of a part of G as shown in Figure 8.

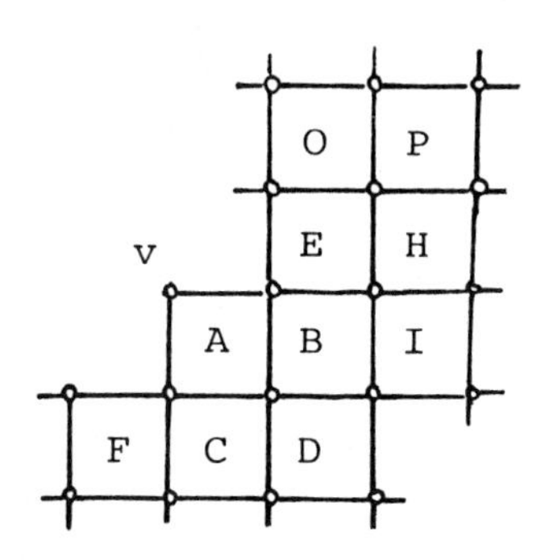

Figure 8.

If $G \supset B, C$, then $G - v$ has a P_3-factor by induction. Hence we may assume that $G \not\supset C$. Moreover, if $G \not\supset B, C$ and $G \supset D$, then $G \not\supset E, F$ which implies that G has a v-semifactor by induction.

The proof is completed by considering the following 12 cases and proving that G has a v-semifactor in each case by induction. However, since the proof of each case is long and tedious we only list the cases:

Case 1. $B, D \not\subset G$

Case 2. $B \subset G, D \not\subset G, F \subset G, E \subset G$

Case 3. $B \subset G, D \not\subset G, F \subset G, E \not\subset G$

Case 4. $B \subset G, D \not\subset G, F \not\subset G, E \subset G$

Case 5. $B \subset G, D \not\subset G, F \not\subset G, E \not\subset G$

Case 6. $B \subset G, D \subset G, E \not\subset G,$

Case 7 $B \subset G, D \subset G, E \subset G, O \subset G, I \subset G$

Case 8. $B \subset G$, $D \subset G$, $E \subset G$, $O \subset G$, $I \not\subset G$

Case 9. $B \subset G$, $D \subset G$, $E \subset G$, $O \not\subset G$, $P \subset G$, $I \subset G$

Case 10. $B \subset G$, $D \subset G$, $E \subset G$, $O \not\subset G$, $P \subset G$, $I \not\subset G$

Case 11. $B \subset G$, $D \subset G$, $E \subset G$, $O \not\subset G$, $P \not\subset G$, $H \subset G$.

Case 12. $B \subset G$, $D \subset G$, $E \subset G$, $O \not\subset G$, $P \not\subset G$, $H \not\subset G$. $\square$

5. APPLICATIONS OF P_3-FACTORS FOR TRIOMINOS TILING PROBLEMS

We remove an arbitrary number of squares from an $m \times n$ chessboard so that the remaining part is connected, and call it a defective board or more briefly a d-board. Two unit squares of a defective chessboard are adjacent if they have a common edge. A d-board B is said to be tough if every pair of adjacent unit squares is contained in a 2×2 subsquare of B. A tough d-board is illustrated in Figure 9.

Figure 9. A tough d-board (black part)

There are exactly two kinds of triominos which have different shape called Tic (Figure 10a) and El (Figure 10b).

Figure 10. Tic and El

By *tiling* a d-board B with triominos we mean covering each square of B exactly once without parts of the triominos extending over the removed square or the edges of the board (see examples in Figure 11).

Figure 11. Tiling a tough d-board by triominos

A square A of a d-board B is called a corner square if A is adjacent with exactly two squares of B.

The order of a d-board B is the number of unit squares of B.

Theorem 5.1. (I) Every tough defective chessboard of order 0 (mod 3) can be tiled with triominos.
(II) Every tough d-board of order 1 (mod 3) can be tiled with triominos except an arbitrary prescribed corner square.

Proof. For a given tough d-board B, we define the graph G(B) such that the vertices of G(B) represent the unit squares of B and the edges of G(B) represent the adjacency of the two corresponding unit squares of B and call those graphs tough graphs. The graph G(B) corresponding to the tough d-board in Figure 11 is illustrated in Figure 12.

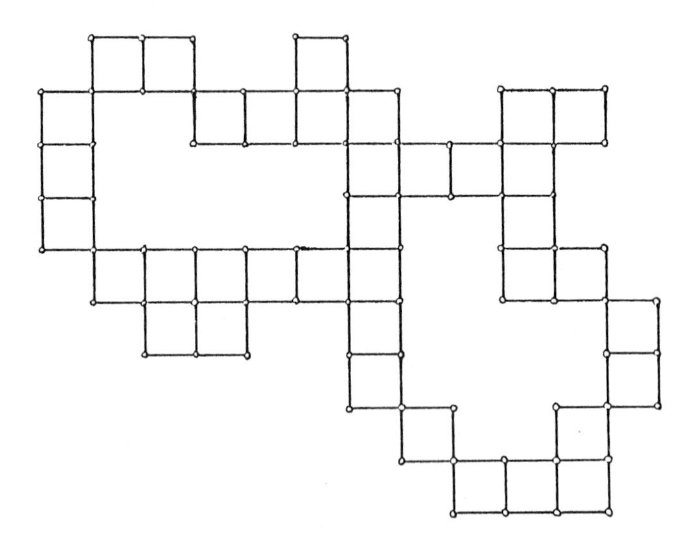

Figure 12. The tough graph corresponding to the tough d-board of Figure 11

It is easily verified that every tough graph is a square graph as defined in Section 5. Applying Theorem 4.2 to G(B), we see that G(B) has a P_3-factor, which implies the theorem. ▯

6. LINEAR ARBORICITY AND STAR DECOMPOSITION INDEX

Let $(G_1, G_2, \ldots, G_n)$ be a set of graphs. If a graph G can be partitioned into edge-disjoing union of $(G_1, G_2, \ldots, G_n)$ -subgraphs of G, then the minimum number of those subgraphs is called the $(G_1, G_2, \ldots, G_n)$-subgraph Decomposition Index of G. In particular, the star decomposition index of G, denoted by $*(G)$, has a star for each G_i, and the $(P_n | n \geqq 2)$-subgraph decomposition index is the same as the linear arboricity, $\Xi(G)$.

The linear arboricity for any r-regular graph G was conjectured to be $\lceil (r+1)/2 \rceil$ in [2]. This has been proved when r = 3,4 in Akiyama, Exoo and Harary [4,5], r = 6 in Tomasta [15], r = 5,6 in Enomoto and Peroche [11].

We now present a few uses of the (P_2, P_3)-factor theorem in linear arboricity problems by showing much shorter proofs than the original ones.

Theorem 6.1. Every cubic graph has linear arboricty 2.

Proof. By Corollary A2, there exist path-factors F of G. Let F' be a path-factor of G having the maximum size among all F's. Denote by H the graph obtained by deleting all edges of F' from G. We claim that H is also a path-factor of G. Suppose that H contains a cycle C, then there are three consecutive vertices

v_1, v_2 and v_3 on C. One of the edges v_1v_2 or v_2v_3 could be added to F' so that either F' + v_1v_2 or F' + v_2v_3 is still a path-factor of G, which contradicts the maximality of F'. □

Theorem 6.2. Every 4-regular graph has linear arboricity 3.

Proof. By Corollary A2, G has a path-factor F. Denote by H the graph obtained from G by deleting all edges of F. Since H can be embedded in some cubic graph, it is union of path-subgraphs F_1 and F_2 by Theorem 6.1. Therefore G is the union of F, F_1, and F_2.

We now turn our attention to star decomposition index.

Theorem 6.3. Let $n \geqq 4$. Then the star decomposition index of the complete graph of order n is $\lceil n/2 \rceil + 1$, i.e.,

$$*(K_n) = \lceil n/2 \rceil + 1$$

Proof. We first show the lower bound $*(K_n) \geqq \lceil n/2 \rceil + 1$ by induction on order n. It suffices to prove that $*(K_{2m-1}) \geqq m+1$ since $*(K_{2m}) \geqq *(K_{2m-1})$ and $\lceil 2m/2 \rceil = \lceil (2m-1)/2 \rceil = m$.

Suppose that $K_{2m-1} = F_1 \cup F_2 \cup \dots \cup F_k$, where each F_i is a star subgraph of K_{2m-1}. If some F_i is a $K(1,2m-2)$, then $K_{2m-1} - F_i = K_{2m-2} \cup K_1 = F_1 \cup \dots \cup F_{i-1} \cup F_{i+1} \cup \dots \cup F_k$ and so $k-1 \geqq *(K_{2m-2}) \geqq m$ by the induction hypothesis. Hence $k \geqq m+1$. Therefore, we may assume that $F_i \neq K(i,2m-2)$ for every i and thus $|E(F_i)| \leqq 2m-3$. Then $|F_1 \cup \dots \cup F_k| \leqq k(2m-3)$. On the other hand, $|E(K_{2m-1})| = (2m-1)m$. Hence we obtain $k \geqq m+1$.

We next show that $*(K_n) \leqq \lceil n/2 \rceil + 1$. It suffices to prove that $*(K_{2m}) \leqq m+1$ since $*(K_{2m-1}) \leqq *(K_{2m})$ and $\lceil 2m/2 \rceil = \lceil (2m-1)/2 \rceil = m$. Let $V(K_{2m}) = \{v_i \mid i=1,2,\ldots, 2m\}$ and put

$$F_\ell = \{v_\ell v_i \mid \ell+1 \leqq i' < \ell+m,\ i \equiv i' \pmod{2m}\}$$

$$\cup \{v_{m+\ell} v_j \mid m+\ell+1 \leqq j' < 2m+\ell,\ j \equiv j' \pmod{2m}\}$$

for $\ell = 1,2,\ldots,m$,

and

$$F_{m+1} = \{v_1 v_{m+1}, v_2 v_{m+2}, \ldots, v_m v_{m+m}\}.$$

Then

$K_{2m} = F_1 \cup F_2 \cup \ldots \cup F_{m+1}$, and thus we have $*(K_{2m}) \leqq m+1$. □

Acknowledgement

It is a pleasure to record our gratitude to K. Ando and H. Enomoto who found the graph in Figure 3 and to D. Avis for his valuable suggestions which have contributed to proof of Theorem 3.1. The first author is grateful to Frank Harary, the Ulam Professor of Mathematics at the University of Colorado in Boulder, Fall 1982, for his kind and generous hospitality at the First Colorado Symposium on Graph Theory which provided him with a productive research atmosphere.

REFERENCES

1. J. Akiyama, Factorization and Linear Arboricity. Doctoral Dissertation, Science Univer. of Tokyo (1982)

2. J. Akiyama, Three Developing Topics in Graph Theory (1980), Science Univ. of Tokyo.
3. J. Akiyama, D. Avis and H. Era, On a {1,2}-factor of a graph, TRU Math. 16(1980) 97-102.
4. J. Akiyama, G. Exoo and F. Harary, Covering and packing in graphs III: Cyclic and acyclic invariants. Math Slovaca 30(1980) 405-417.
5. J. Akiyama, G. Exoo and F. Harary, Covering and packing in graphs IV: Linear arboricity. Networks 11(1981) 69-72.
6. A. Amahashi and M. Kano, On factors with given components. Discrete Math. 42 (1982) 1-6.
7. H. B. Belck, Regulare Faktoren von Graphen. J. Reine Angew. Math. 188(1950) 228.
8. C. Berge, Graphs et Hypergraphs, Dunod, Paris (1970).
9. C. Berge, Some common properties for regularizable graphs, edge-critical graphs and B-graphs. Springer Lecture Notes in Computer Science, 108 (1981) 108-123.
10. G. Cornuejols and W. Pulleyblank, A matching problem with side conditions. Discrete Math. 29(1980) 135-159.
11. H. Enomoto and C. Peroche, The linear arboricity of 5-regular graphs. J. Graph Theory (to appear).
12. A. Hajnal, A theorem on k-saturated graphs. J. London Math. Soc. 15(1947) 107-111.
13. F. Harary, Graph Theory, Addison-Wesley, Reading, Mass. (1969).
14. J. Plesnik, Connectivity of regular graphs and the existence of 1-factors, Mat. Capo. 22(1972) No. 4, 310-318.

15. P. Tomasta, Note on linear arboricity, Math. Slovaca 32(1982) 239-242.

16. W. T. Tutte, The factorization of linear graphs. J. London Math. Soc. 22(1947) 107-111.

17. W. T. Tutte, The factors of graphs. Canad. J. Math. 4(1952) 314-328.

THE WREATH PRODUCT OF GRAPHS

Marlow Anderson
The Colorado College
Colorado Springs, Colorado 80903

Marc Lipman
Indiana University-
Purdue University at Ft. Wayne
Ft. Wayne, Indiana 46805

ABSTRACT. A new graph product called the wreath product is defined, whose vertex set is the cartesian product of the vertices of the two factors, and which lies between the cartesian product and the composition of graphs. It is the smallest such product whose automorphism group contains the wreath product of the automorphism groups of the factors. For large classes of graphs, equality is obtained. Partial results are adduced toward the conjecture that if the first factor is of class one, then the wreath product is.

1. THE DEFINITION OF WREATH PRODUCT

For a graph G we denote its vertex set and edge sets by VG and EG, respectively. We consider only simple graphs, and denote an edge between two vertices by juxtaposition. We adopt the standard notations K_n, P_n and C_n for the complete graph, path and cycle on n vertices, respectively. The complement of a graph G is written $\bar{G}$, and the adjacency matrix of G is A(G). All groups are written with script letters; $\mathcal{S}_n$ is a symmetric group on n objects and $\mathcal{Z}_n$ is the cyclic group of order n. In particular, the automorphism group of a graph G is denoted by $\mathcal{G}$ or $\mathcal{A}(G)$.

An association which assigns to each ordered pair of graphs G and H a graph $G \circ H$ on the vertex set $VG \times VH$ is called a <u>boolean operation</u> by Harary and Wilcox [4]; we shall follow the notation and terminology used there. Some work has been done relating $\mathcal{A}(G \circ H)$ to $\mathcal{G}$ and $\mathcal{H}$ ([2], [3, Chap. 12], [6], [7]). Our object is to define a new boolean operation ρ so that $\mathcal{A}(G \rho H)$ always contains the wreath product of $\mathcal{G}$ and $\mathcal{H}$, and such that ρ is in some sense the smallest such boolean operation.

This new operation will lie between the previously defined cartesian product and composition. The edge set of the <u>cartesian product</u> $G \times H$ is

$$E = \{(g,h)(g,h') : hh' \in EH\} \cup \{(g,h)(g',h) : gg' \in EG\};$$

the edge set of the <u>composition</u> G[H] (first defined in [2]) is

$$E = \{(g,h)(g,h') : hh' \in EH\} \cup \{(g,h)(g',h') : gg' \in EG\}.$$

The <u>wreath product</u> $G \rho H$ has edge set

$$E = \{(g,h)(g,h') : hh' \in EH\} \cup \{(g,h)(g',h') : gg' \in EG \text{ and there exists } \alpha \in \mathcal{H} \text{ so that } h\alpha = h'\}.$$

It is then obvious that

$$G \times H \subseteq G \rho H \subseteq G[H].$$

Furthermore, $G \rho H = G \times H$ exactly when $\mathcal{H}$ is trivial, and $G \rho H = G[H]$ exactly when $\mathcal{H}$ is transitive on H. As an example of the latter case, $K_n \rho K_m = K_{mn}$, and so $G \rho H$ is complete exactly when G and H are.

These inclusions can be further illustrated by examining the adjacency matrices of these three graphs. Denote by I_m the $m \times m$ identity matrix, and by J_n the $n \times n$ matrix consisting of all ones; let $*$ be matrix tensor product and $\oplus$ matrix addition modulo 2. Then, if $|VG| = m$ and $|VH| = n$, and P is the $m \times n$ matrix $\{p_{ij}\}$ with

$$p_{ij} = \begin{cases} 1 & \text{if } h_i, h_j \text{ belong to the same orbit of } \mathcal{H} \\ 0 & \text{otherwise,} \end{cases}$$

we have

$$A(G \times H) = (A(G) * I_n) \oplus (I_m * A(H)) \ ,$$

$$A(G \rho H) = (A(G) * P) \oplus (I_m * A(H))$$

and

$$A(G[H]) = (A(G) * J_n) \oplus (I_m * A(H)) \ .$$

These matrices for $A(G \times H)$ and $A(G[H])$ were mentioned in [4].

Now $G \rho H$ can be decomposed in a natural way into copies of H. Let $V = V(G \rho H)$ and for $g \in VG$, let $H_g = \{g\} \times VH$. Then $V = \cup \{H_g : g \in VG\}$, and the induced subgraph of $G \rho H$ on H_g is isomorphic to H.

We will now count the edges and vertex degrees in $G \rho H$. Let $\{\mathcal{O}_i\}_{i=1}^{r}$ be the set of orbits of $\mathcal{H}$ acting on H, and suppose that each $h \in \mathcal{O}_i$ has degree e_i. For each copy H_g of H in $G \rho H$, there are $|EH|$ edges, while for each edge of G and each orbit $\mathcal{O}_i$ there are $|\mathcal{O}_i|^2$ edges, making

$$|VG|\,|EG| + |EG| \sum_{i=1}^{r} |\mathcal{O}_i|^2$$

edges altogether in $G \rho H$. If $g \in VG$ has degree d in G and $h \in \mathcal{O}_i$, then clearly the degree of vertex (g,h) in $G \rho H$ is $d|\mathcal{O}_i| + e_i$.

We next count the components of $G \rho H$. Denote by $\omega(K)$ the number of components of graph K. Let n be the number of isolated vertices of G, and m the number of isomorphism classes of components of H.

Then

$$\omega(G \rho H) = n\,\omega(H) + (\omega(G) - n)m$$

In particular, if G is nontrivial, then $G \rho H$ is connected if and only if G is connected and all components of H are isomorphic.

We shall now conclude this section by proving the theorems which justify calling $G \rho H$ the wreath product of graphs.

Theorem 1.1. $\mathcal{G} \rho \mathcal{H}$ is a subgroup of $\mathcal{A}(G \rho H)$.

Proof. Suppose that $|VG| = n$. Then $\mathcal{G}$ acts on a set of n elements and we mean by $\mathcal{G} \rho \mathcal{H}$ the group of $|\mathcal{G}| \cdot |\mathcal{H}|^n$ elements with typical element

$$\Phi = (\sigma;\ \tau_1, \ldots, \tau_n), \text{ and } \sigma \in \mathcal{G},\ \tau_i \in \mathcal{H}.$$

We define Φ acting on $G \rho H$ as follows.

Let $(g_i, h) \in V(G \rho H)$ and suppose that $g_i\sigma = g_j$;

Then

$$(g_i, h)\Phi = (g_i, h)(\sigma;\ \tau_1, \ldots, \tau_n) = (g_j, h\tau_j).$$

To show that $\Phi \in \mathcal{A}(G \rho H)$ we must check that it preserves edges. First suppose that $hh' \in EH$ and so $(g_i,h)(g_i,h') \in E(G \rho H) = E$. Then $(g_i,h)\Phi = (g_j, h\tau_j)$ and $(g_i,h')\Phi = (g_j, h'\tau_j)$. Since $hh' \in EH$, $(h\tau_j)(h'\tau_j) \in EH$ and so

$$(g_i,h)\Phi\ (g_i,h')\Phi \in E.$$

On the other hand, if $g_ig_k \in EG$ and $h = h'$ for $\alpha \in \mathcal{H}$, then $(g_i,h)(g_k,h') \in E$. Now $(g_i,h)\Phi = (g_j, h\tau_j)$ and

$(g_k,h')\Phi = (g_\ell, h'\tau_\ell)$, where $g_k\sigma = g_\ell$. But $g_jg_\ell \in EG$ and $(h\tau_j)\tau_j^{-1}\alpha\tau_\ell = h'\tau_\ell$ and so

$$(g_i,h)\Phi(g_k,h')\Phi \in E.$$

We now must show that this embedding preserves multiplication, that is, for all (g_i,h) and for all Γ, $\Delta \in \mathcal{G} \rho \mathcal{H}$, $(g_i,h)(\Gamma\Delta) = ((g_i,h)\Gamma)\Delta$. If $\Gamma = (\sigma; \tau_1,\ldots,\tau_n)$ and $\Delta = (\nu; \phi_1,\ldots,\phi_n)$, then

$$\Gamma\Delta = (\sigma\nu; \tau_{1\nu^{-1}}\phi_1, \tau_{2\nu^{-1}}\phi_2, \ldots \tau_{n\nu^{-1}}\phi_n).$$

Suppose that $g_i\sigma = g_j$ and $g_j\nu = g_k$. Then $(g_i,h)\Gamma = (g_i,h\tau_j)$ and $(g_j,h\tau_j)\Delta = (g_k,h\tau_j\phi_k)$. But $j = k\nu^{-1}$ and so we have what is necessary. □

Theorem 1.2. Let K be a graph with vertex set $VG \times VH$ which contains $G \times H$. If $\mathcal{G} \rho \mathcal{H}$ is a permutation subgroup of $\mathcal{K}$, then $G \rho H$ is a subgraph of K.

<u>Proof</u>. Suppose that $g_1g_2 \in EG$ and there exists $\alpha \in \mathcal{H}$ with $h\alpha = h'$; that is, $(g_1,h)(g_2,h') \in E(G \rho H)$. Now $(g_1,h)(g_2,h) \in E(G \times H) \subseteq E(K)$, while

$$\Phi = (\iota; \iota, \alpha, \iota, \ldots, \iota) \in \mathcal{G} \rho \mathcal{H} \subseteq \mathcal{K}$$

Then $(g_1,h)\Phi = (g_1,h)$ and $(g_2,h)\Phi = (g_2,h')$ and so

$$(g_1,h)(g_2,h') \in E(K).$$ □

2. THE AUTOMORPHISM GROUP OF THE WREATH PRODUCT

We shall now consider when the automorphism group of the wreath product of two graphs is equal to the wreath product of their automorphism groups. This is not always the case, but does hold for large classes of graphs. We do not have a necessary and sufficient condition. The following lemma is obvious but useful.

Lemma 2.1. If $\phi \in \mathcal{A}(G \rho H)$ and for all $g \in G$ there exists $g' \in G$ with $\phi(H_g) = H_{g'}$ (that is, ϕ preserves copies of H in G ρ H), then $\phi \in \mathcal{G} \rho \mathcal{H}$. ▯

Now Sabidussi [6,7] has obtained necessary and sufficient conditions that $\mathcal{A}(G[H])$ be $\mathcal{G} \rho \mathcal{H}$. Since G[H] and G ρ H are the same when $\mathcal{H}$ is transitive, this result does answer our question in this case. The Sabidussi Theorem is as follows:

Theorem 2.2. $\mathcal{A}(G[H])$ is equal to $\mathcal{G} \rho \mathcal{H}$ exactly when these two conditions hold:

(1) if two distinct vertices of G have the same deleted neighborhoods, then H is connected; and

(2) if two distinct vertices of G have the same closed neighborhoods, then $\bar{H}$ is connected.

It is easily checked that if $\mathcal{A}(G \rho H) = \mathcal{G} \rho \mathcal{H}$, then Sabidussi's proof [6] of the necessity of his two conditions still holds. However, his conditions are no longer sufficient. For let H be the path P_3 with three vertices and G be the graph $K_4 - e$. Using Lemma 2.1 it can be checked directly that $\mathcal{A}(G \rho H) = \mathcal{G} \rho \mathcal{H}$. However, the two vertices of degree three have the same closed neighborhood and so condition (2) above fails, so that

$$\mathcal{A}(G[H]) \subset \mathcal{A}(G \rho H) = \mathcal{G} \rho \mathcal{H} .$$

Suppose that G is the graph $K_4 - e$, and H is a graph with $\mathcal{H}$ transitive and $\bar{H}$ disconnected. Then the Sabidussi result implies that $\mathcal{A}(G \rho H) \subset \mathcal{G} \rho \mathcal{H}$. However, it is possible to obtain $\mathcal{G} \rho \mathcal{H}$ exactly from G ρ H and G ρ $\bar{H}$, in the following sense.

Theorem 2.3. Consider $\mathcal{A}(G \rho H)$ and $\mathcal{A}(G \rho \bar{H})$ as subgroups of the symmetric group on $VG \times VH$. Then

$$\mathcal{A}(G \rho H) \cap \mathcal{A}(G \rho \bar{H}) = \mathcal{G} \rho \mathcal{H} .$$

Proof. That the intersection above contains $\mathcal{G} \rho \mathcal{H}$ is clear by Theorem 1.1, since $\mathcal{A}(\bar{H}) = \mathcal{A}(H) = \mathcal{H}$. Since at least one of H and $\bar{H}$ is connected, we may suppose that H is. Now, if $\gamma \in \mathcal{A}(G \rho H) - \mathcal{G} \rho \mathcal{H}$, then by Lemma 2.1 there exist vertices x and y in the same copy H_g of H in $G \rho H$ with $x\gamma$ and $y\gamma$ in different copies. Since H is connected, there is a path in H_g connecting x and y, and we can in fact assume that this path is a single edge xy. Now the edge $x\gamma y\gamma$ in $G \rho H$ is between copies of H, and so the H-coordinates of $x\gamma$ and $y\gamma$ are in the same orbit of $\mathcal{H}$. But then $x\gamma y\gamma$ is also an edge in $G \rho \bar{H}$ while xy is not, which means that $\gamma \notin \mathcal{A}(G \rho \bar{H})$. Result follows. ▯

Note that though the intersection of $\mathcal{A}(G \rho H)$ and $\mathcal{A}(G \rho \bar{H})$ is $\mathcal{G} \rho \mathcal{H}$ neither group alone need be. For let G be the graph $K_4 - e$ and K_n the complete graph on n vertices ($n > 1$). Then since $\mathcal{A}(K_n)$ is transitive, by Theorem 2.2, both $\mathcal{A}(G \rho K_n)$ and $\mathcal{A}(G \rho \bar{K}_n)$ properly contain $\mathcal{G} \rho \mathcal{A}(K_n)$.

We shall conclude this section by providing some examples where $\mathcal{G} \rho \mathcal{H}$ is the entire automorphism group. The first is a corollary of the Sabidussi Theorem.

Corollary 2.4. The wreath product of two cycles satisfies the conditions:

$$\mathcal{S}_3 \rho \mathcal{S}_3 = \mathcal{C}_3 \rho \mathcal{C}_3 \subset \mathcal{A}(C_3 \rho C_3) = \mathcal{A}(K_9) = \mathcal{S}_9 ,$$

and if at least one of n and m is not 3, then

$$\mathcal{C}_n \rho \mathcal{C}_m = \mathcal{A}(C_n \rho C_m) . \qquad ▯$$

Theorem 2.5. The wreath product of two paths satisfies:

$$\mathcal{Z}_2 \,\rho\, \mathcal{Z}_2 = \mathcal{P}_2 \,\rho\, \mathcal{P}_2 \subset \mathcal{A}(P_2 \,\rho\, P_2) = \mathcal{A}(K_4) = \mathcal{S}_4,$$
$$\mathcal{Z}_2 \,\rho\, \mathcal{Z}_2 = \mathcal{P}_2 \,\rho\, \mathcal{P}_3 \subset \mathcal{A}(P_2 \,\rho\, P_3) = \mathcal{A}(K_{3,3}),$$

and for every other m and n ,

$$\mathcal{P}_n \,\rho\, \mathcal{P}_m = \mathcal{A}(P_n \,\rho\, P_m).$$

Proof. The results regarding $P_2 \,\rho\, P_2$ are obvious, and so are those regarding $P_2 \,\rho\, P_3$, since the automorphism group of the complete bipartite graph $K_{3,3}$ has 72 elements.

First let n be 2. There are then two cases, depending on whether m is even or odd, illustrated for m = 4 and m = 5 in Figure 1.

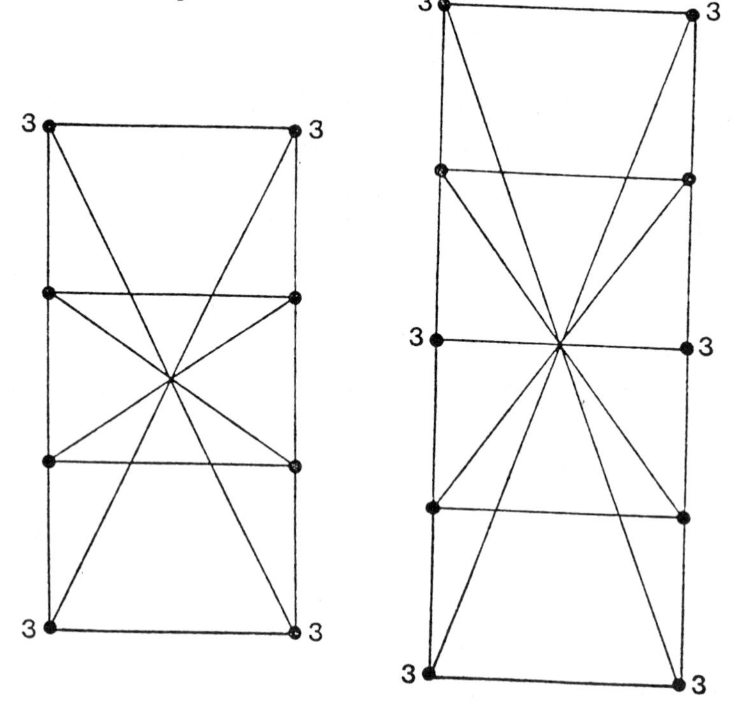

Figure 1.

When m is even, the corners are the only vertices with degree 3. Since each corner is not adjacent to the corner at the opposite end of its copy of P_m but is adjacent to the other two, specifying the image of one corner under an automorphism forces the corners in a copy of P_m to be mapped into another copy of P_m. But once the images of the corners are so determined, this forces the copies of P_m to be preserved by the automorphism. When m is odd, there are 6 vertices of degree 3, but only the corners are adjacent to two other degree 3 vertices. Thus the same argument applies, since a middle vertex of degree 3 is of distance $(m-1)/2$ from the corners in its copy of P_m and $(m+1)/2$ from the others.

When $n > 2$, the arguments are similar. When m is even, the $m \times n$ matrix of vertex degrees is of the form

$$\begin{bmatrix} 3 & 5 & \cdots & 5 & 3 \\ 4 & 6 & \cdots & 6 & 4 \\ \vdots & \vdots & & \vdots & \vdots \\ 4 & 6 & \cdots & 6 & 4 \\ 3 & 5 & \cdots & 5 & 3 \end{bmatrix}$$

The corners are the only vertices of degree 3, and since corners in the same copy of P_m are connected by a path through vertices of degree 4, corners in the same copy of P_m go to the same copy under any automorphism. The same sort of argument is then applied to the end vertices of degree 5, finally allowing us to apply Lemma 2.1. When m is odd, the $m \times n$ matrix of vertex degrees is of the form

$$\begin{bmatrix} 3 & 5 & \cdots & 5 & 3 \\ 4 & 6 & \cdots & 6 & 4 \\ \vdots & \vdots & & \vdots & \vdots \\ 4 & 6 & \cdots & 6 & 4 \\ 3 & 4 & \cdots & 4 & 3 \\ 4 & 6 & \cdots & 6 & 4 \\ \vdots & \vdots & & \vdots & \vdots \\ 4 & 6 & \cdots & 6 & 4 \\ 3 & 5 & \cdots & 5 & 3 \end{bmatrix}$$

and similar reasoning is applicable. ▯

Finally, for $n \geq 3$ let W_n be the wheel consisting of a cycle C_n with a vertex adjoined adjacent to all the other vertices. Then we have the following.

Theorem 2.6.

$$\mathcal{S}_4 \,\rho\, \mathcal{S}_4 = \mathcal{W}_3 \,\rho\, \mathcal{W}_3 \subset \mathcal{A}(W_3 \,\rho\, W_3) = \mathcal{A}(K_8) = \mathcal{S}_{16}.$$

If $n > 3$ and $m \geq 3$, then

$$\mathcal{W}_n \,\rho\, \mathcal{W}_m = \mathcal{A}(W_n \,\rho\, W_m) .$$

<u>Proof</u>. Since $W_3 = K_4$, the first assertion is obvious. If $n > 3$ and $m = 3$, then Theorem 2.2 applies.

For $n, m > 3$, we count the vertex degrees as follows. Denote by (c,c), (c,o), (o,c) and (o,o) vertices in $W_n \rho W_m$ with W_n and W_m components as center vertices (c) or outer vertices (o), respectively, Then

$$\text{degree }(c,c) = n + m\,,$$
$$\text{degree }(c,o) = 3 + nm\,,$$
$$\text{degree }(o,c) = 3 + m\,, \text{ and}$$
$$\text{degree }(o,o) = 3 + 3m\,.$$

By making all possible comparisons and solving for n, it is easy to see that these four degrees are distinct for all $n, m > 3$, except possibly for (c,c) and (o,o) vertices. But the (c,c) vertex is not adjacent to any other vertex of the same degree while (o,o) vertices always are. Consequently, any automorphism of $W_n \rho W_m$ leaves (c,c) fixed. By considering the vertices adjacent to it, it is clear that the center copy of W_n is then mapped to itself. Since the set of (o,c) vertices is mapped onto itself, adjacency considerations now lead us to conclude that copies of W_m are preserved, and so by Lemma 2.1, $\mathcal{A}(W_n \rho W_m) = \mathcal{W}_n \rho \mathcal{W}_m$. ▯

3. EDGE COLORINGS OF WREATH PRODUCTS

We now consider edge colorings of $G \rho H$. As in [3], we denote the edge chromatic number of a graph K by $\chi'(K)$ and the maximum degree of K by $\Delta(K)$. Then K is said to be of class 1 if $\chi'(K) = \Delta(K)$ and of class 2 if $\chi'(K) = \Delta(K) + 1$ (see [1, p. 91]).

We have the following edge coloring results for the cartesian product (see [1, 6.2.6] or [4]) and the composition.

Theorem 3.1. If G is of class 1, then $G \times H$ is of class 1. ▯

Theorem 3.2. If G is of class 1, then G[H] is of class 1.

<u>Proof</u>. Now, $\Delta(G[H]) = \Delta(H) + \Delta(G) \cdot |VH|$. Color each copy of H in G[H] in the same way with $\chi = \chi_1(H)$ colors $\{k_1,\ldots,k_\chi\}$. Color G with $\chi_1(G) = \Delta(G)$ colors $\{c^1,\ldots,c^{\Delta(G)}\}$. Each edge uv in G corresponds to a complete bipartite subgraph of G[H] on the bipartition H_u, H_v. Suppose uv has color c^i. The edges between H_u and H_v can be colored with colors $\{c_1^i,\ldots,c_{|VH|}^i\}$ so that the edges (u,h)(v,h) have color c_1^i.

If $\chi = \chi_1(H) = \Delta(H)$, then a total of $\Delta(G[H])$ colors has been used. If $\chi = \chi_1(H) = \Delta(H) + 1$, then for each $h \in VH$, there is at least one k_i that is not used. Replace every edge (u,h)(v,h) colored c_1^1 with an unused k_i. Thus G[H] is colored with $\Delta(G[H])$ colors. ▯

Since $G \times H = G \rho H$ when $\boldsymbol{H}$ is trivial and $G[H] = G \rho H$ when $\boldsymbol{H}$ is transitive it is reasonable to make the following:

<u>Conjecture</u>. If G is of class 1, then $G \rho H$ is of class 1.

Before examining further evidence for the conjecture, it is worthwhile to note that it cannot be strengthened. More precisely, for any triple of classes

$$(\text{class of } G, \text{ class of } H, \text{ class of } G \rho H) = (2, i, j)$$

where i, j = 1,2, there exist graphs G and H which satisfy the triple, as the following examples reveal.

Example 3.1. (2, 1, 1). Let $G = K_{2n+1}$ and $H = K_{2m}$. Then $G \rho H = K_{4mn+2m}$ is complete on an even number of vertices. Therefore, G is of class 2 while H and $G \rho H$ are of class 1.

Example 3.2. (2, 1, 2). Let $G = K_3$ and $H = P_3$. Then $\Delta(G \rho H) = 5$, and it has 6 vertices of degree 5 and 3 vertices of degree 4. Thus $G \rho H$ has 21 edges, no more than $A = \lfloor 9/2 \rfloor$ of which are independent. Hence each color class contains at most 4 edges and so $\chi_1(G \rho H) \geq 6$. Thus G is of class 2, H is of class 1 and $G \rho H$ is of class 2.

Example 3.3. (2, 2, 1). Let $G = K_3$ and $H = K_5 - e$. Then H is isomorphic to $K_3 + \bar{K}_2$ (here + means join; see [3]) and *H* has 2 orbits, those vertices "in" K_3 and those "in" $\bar{K}_2$. Let $VG = \{\alpha, \beta, \gamma\}$ and $VH = \{a,b,c,d,e\}$, with de not an edge of H. Then, the edge set of $G \rho H$ can be decomposed into 5 disjoint subsets: edges of H_ω, with $\omega = \alpha, \beta, \gamma$; edges between copies of $\bar{K}_2$; and edges between copies of K_3. Now, $\Delta(G \rho H) = 10$, and Figures 2, 3, and 4 display a 10-coloring of the edges of $G \rho H$ displayed by the above decomposition. Thus, $G \rho H$ is of class 1. But H has 9 edges and at most $2 = \lfloor 5/2 \rfloor$ edges are independent. Thus $\chi_1(H) \geq 5$ and so H is of class 2, as is G.

Example 3.4. (2, 2, 2). It is well known that a regular graph on an odd number of vertices is of class 2. Thus, the following graphs are all of class 2: C_{2n+1}, C_{2m+1}, $C_{2n+1} \rho C_{2m+1}$; K_{2m+1}, $K_{2n+1} \rho K_{2m+1} = K_{4mn+2(m+n)+1}$.

We do have the following special case of the conjecture:

Theorem 3.3. Let G be of class 1, and suppose that H has the property that a vertex in the largest isomorphism class of vertices in H has maximum degree in H. Then $G \rho H$ is of class 1.

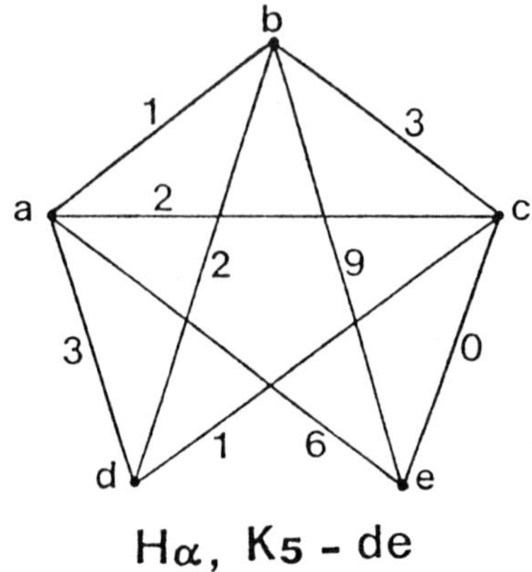

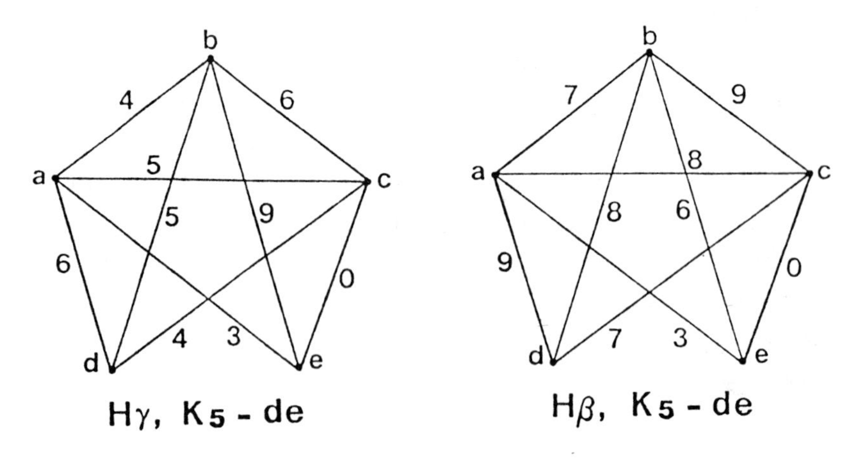

Figure 2.

Proof. Let j be the number of vertices in the largest isomorphism class of vertices in H. Then the maximum degree $\Delta(G \rho H)$ is $\Delta(G) \cdot j + \Delta(H)$. Color G with colors $\{c^1, \ldots, c^{\Delta(G)}\}$ and then use this coloring as in proof of Theorem 3.2 to color the edges of $G \rho H$ which span different copies of H, with $\Delta(G) \cdot j$ colors, $\{c_1^1, \ldots, c_1^{\Delta(G)}, \ldots, c_j^{\Delta(G)}\}$. This can be done so that color c_1^i occurs only on edges of the form $(x,h)(y,h)$. Color each copy of

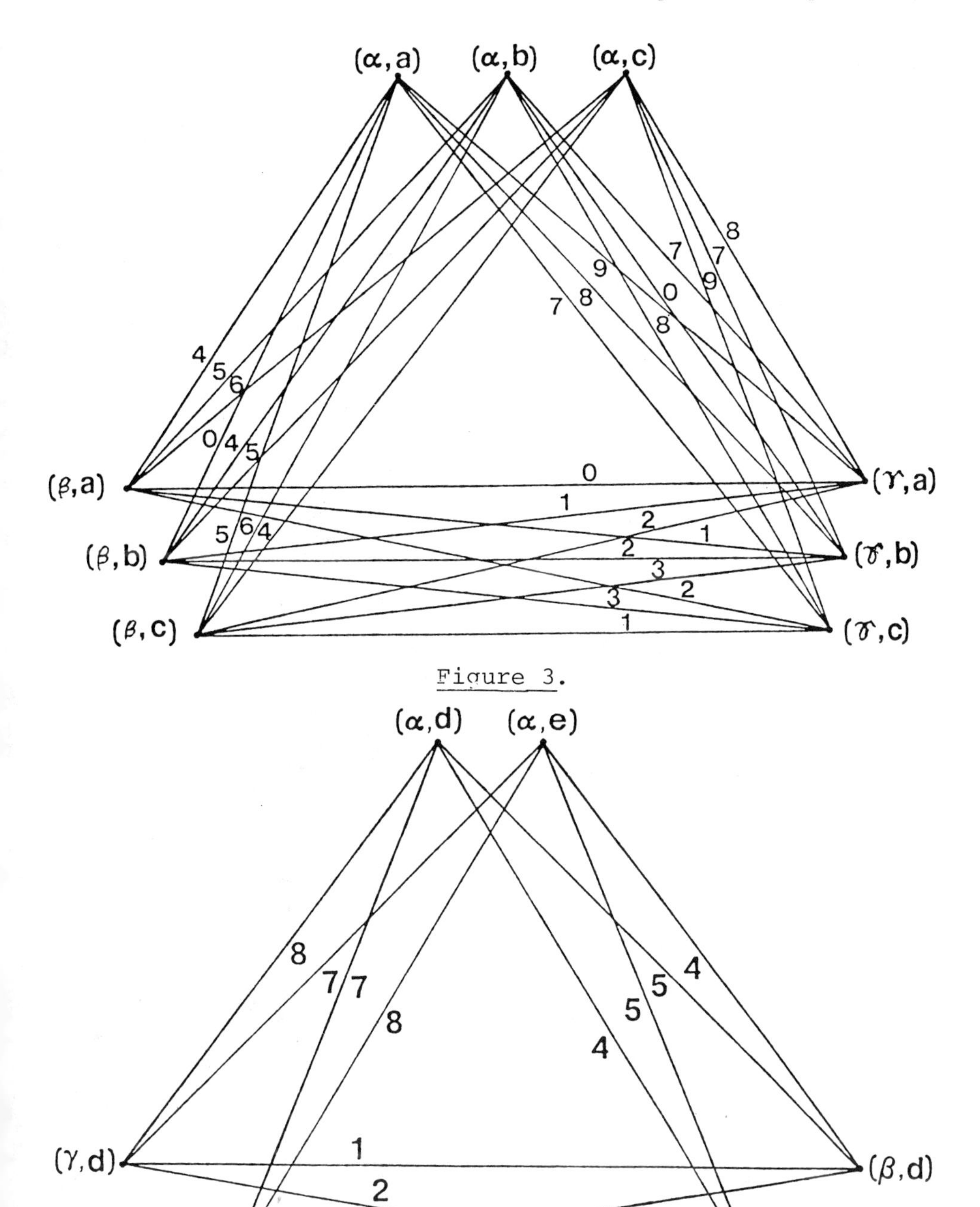

Figure 3.

Figure 4.

H identically with $\chi_1(H)$ colors. If $\chi_1(H) = \Delta(H)$, we are done. If $\chi_1(H) = \Delta(H) + 1$, then at each vertex of H there is an unused color. We can then replace all edges colored c_1^1 with the appropriate unused color, and the coloring is complete. □

This theorem provides large classes of graphs H for which the conjecture holds:

(a) $H = K_n - \mathcal{E}$, where $\mathcal{E}$ is a set of less than $\lfloor n/2 \rfloor$ disjoint edges;

(b) $H = H_1 + H_2$, where H_i is a vertex transitive graph of degree d_i on n_i vertices, and $d_1 + n_2 \geq d_2 + n_1$. For example: $K_n + C_m$, $n \geq 2$, $m \geq 3$.

(c) $H = P_n$, where $n \neq 3$.

Although the theorem does not apply, one can also check that $G \rho P_3$ is of class 1 if G is. In fact, one can use similar methods to show that $G \rho P_n$ is of class 1 for any graph G, if $n > 3$. However, Example 3.2 provides an example to show that $G \rho P_3$ need not be class 1 if G is class 2. Furthermore, $C_5 \rho P_2$ is of class 1, but no proper 5-coloring of it comes from or yields an optimal coloring of C_5. Thus, the question of graphs $G \rho P_2$ where G is of class 2 is probably difficult.

The conjecture can also be shown to be true for many graphs H which do not satisfy the criterion of Theorem 3.3. In particular, one can prove that if G is of class 1 and W_m is the wheel as defined in Section 2, with $m \geq 3$, then $G \rho W_m$ is of class 1.

REFERENCES

1. J. A. Bondy and U. Murty, Graph Theory with Applications. North Holland, New York, (1980). MR 54 #117.
2. F. Harary, On the group of the compositon of two graphs. Duke Math. J. 26 (1959) 29-24. MR22 #1523.
3. F. Harary, Graph Theory. Addison Wesley, Reading, Mass., (1969). MR 41 #1566.
4. F. Harary and G. Wilcox, Boolean operations on graphs. Math. Scand. 20 (1967) 41-54. MR 35 #2775.
5. E. S. Mahamoodian, On edge-colorability of cartesian products of graphs. Canad. Math. Bull. 24 (1984) 107-108. MR 82e:05068.
6. G. Sabidussi, The composition of graphs. Duke Math. J. 26 (1959) 693-696. MR 22 #1524.
7. G. Sabidussi, The lexicographic product of graphs. Duke Math. J. 28 (1961) 573-578. MR 26 #4345.

PACKINGS OF COMPLETE BIPARTITE GRAPHS

Lowell W. Beineke

Department of Mathematical Sciences
Indiana University-Purdue University at Fort Wayne
Fort Wayne, IN 46805

ABSTRACT. A formula is found for the maximum number of copies of the complete bipartite graph $K_{r,s}$ which can be packed into $K_{m,n}$.

Combinatorial packing problems have been found to be intriguing to many mathematicians. The most famous of these problems is probably that of packing spheres, and that has led to a considerable theory (see, for example, the book Packing and Covering by C.A. Rogers [6] or the recent article by N.J.A. Sloane [7]). A second famous packing problem (actually a decomposition problem) is that of Steiner systems. Somewhat surprisingly, these two problems have close connections; and for an interesting and elementary discussion of both, see Anderson [1, Chapter 7].

The simplest case of the Steiner problem is, in terms of graphs, to determine those values of n for which the complete graph K_n can be decomposed into triangles (namely, $n \equiv 1$ or $3 \pmod 6$). As an extension of this result, Guy [5] showed that the maximum number of edge-disjoint triangles which can be packed into K_n is $\left[\frac{1}{3} n \left[\frac{1}{2}(n-1)\right]\right]$ (the brackets denoting the greatest integer function) unless $p \equiv 5 \pmod 6$, in which case it is one less. (For a discussion of a number of related problems, see Guy [5].)

Chartrand, Geller, and Hedetniemi [4] considered the problem of packing cycles into graphs, and, using Guy's result, showed that the maximum number in K_n is always $\left[\frac{1}{3}n\,[\frac{1}{2}(n-1)]\right]$. They also solved this problem for every complete bipartite graph $K_{m,n}$, showing that the maximum number is $\min\left\{\left[\frac{m}{2}[\frac{n}{2}]\right], \left[\frac{n}{2}[\frac{m}{2}]\right]\right\}$; and that an optimal packing is always realizable with 4-cycles. Since a 4-cycle is isomorphic to the complete bipartite graph $K_{2,2}$, their problem suggests the larger problem that we consider here, namely, packing other complete bipartite graphs into $K_{m,n}$.

In point of fact, however, our interest in the problem originated in a setting different from this. The "coarseness" of a graph in the maximum number of nonplanar graphs in a packing (in contrast to the "thickness," the minimum number of planar graphs in a covering), and the maximum number of subgraphs in a packing with $K_{3,3}$'s is a lower bound for the coarseness. In investigating the coarseness of $K_{m,n}$, Beineke and Guy [3] showed that this number is $\min\left\{\left[\frac{m}{3}[\frac{n}{3}]\right], \left[\frac{n}{3}[\frac{m}{3}]\right]\right\}$ (and in many cases this is the coarseness of $K_{m,n}$). Later, it was announced that this result could be extended to arbitrary $K_{r,r}$ packings of $K_{m,n}$ [2]. However, what was thought to be a proof had a gap in certain cases. Fortunately, this gap could be filled, and we have been able not only to complete the proof, but to extend the result to packings with $K_{r,s}$.

Formally, the general definition of packing one graph with another is the following: a *packing of* G *with* H is a collection of edge-disjoint subgraphs of G all of which are isomorphic to H. These copies of H are called the *elements* of the packing, and the *packing number* $\mu_H(G)$ is the maximum number of elements in any packing of G with H. (Clearly these definitions can be extended to packing a graph with graphs in some family.)

The specific problem of concern here is determining the *complete bipartite packing number* $\mu := \mu_{r,s}(m,n)$ for packing $K_{m,n}$ with $K_{r,s}$, with partite sets preserved. (This last condition means that in any element of the packing, the r vertices in $K_{r,s}$ come from the set of m in $K_{m,n}$ and the s from the set of n.) For convenience, we shall refer to the vertices in the first partite set of a complete bipartite graph as red and those in the second as blue.

We now derive an upper bound for μ. Since each red vertex has valency n in $K_{m,n}$ and valency s in $K_{r,s}$, it can appear in at most $\left[\frac{n}{s}\right]$ elements of a packing. Hence, there are at most $m\left[\frac{n}{s}\right]$ occurrences of red vertices in the entire packing, and since each element has r red vertices, there can be at most $\left[\frac{m}{r}\left[\frac{n}{s}\right]\right]$ elements in the packing. Likewise, there can be no more than $\left[\frac{n}{s}\left[\frac{m}{r}\right]\right]$, so that $\mu \le \nu$, where ν is defined as

$$\nu := \min\left\{\left[\frac{m}{r}\left[\frac{n}{s}\right]\right],\ \left[\frac{n}{s}\left[\frac{m}{r}\right]\right]\right\}.$$

Observe that, if r divides m and n divides s, then (and only then) there is a perfect packing (or decomposition) of $K_{m,n}$ with $\frac{mn}{rs}$ graphs $K_{r,s}$ and so $\mu = \nu$ in this simple case.

We now turn to establishing the inequality $\mu \geq \nu$ in general. This will be done constructively in two stages. Although the first construction does not suffice in all cases, we present it separately since it (i) is simpler than the general construction and is the basis for the second, and (ii) is sufficient in the "vast majority" of cases including all of those in which $r = s = 2$ or 3 mentioned earlier.

Let $m = pr + t$ with $0 \leq t < r$ and $n = qs + u$ with $0 \leq u < s$. Then $\nu = pq + \min\left\{[\frac{qt}{r}], [\frac{pu}{s}]\right\}$. We assume, without loss of generality, that $\frac{pu}{s} \leq \frac{qt}{r}$, and we let $a: = [\frac{pu}{s}]$. Furthermore, since $K_{pr,qs}$ is a subgraph of $K_{m,n}$, it follows from the earlier observation (two paragraphs above) that $\mu = \nu$ if $a = 0$. Therefore we assume that $a > 0$ (which implies that both t and u are also positive).

In our construction, we form an array of ν cells, into each of which we will put r red vertices and s blue vertices to form a complete bipartite graph. The array has p full rows of q cells each and a partial $(p+1)$th row with a cells. Within each full row, the q cells are divided into r blocks, as nearly equal as possible; the partial row is divided in the same pattern as far as it goes.

Formally, this is done as follows: The number of cells in the k'th block (for $k = 1,2,\ldots,r$) of each full row is

$$N_k: = \left[\frac{qk}{r}\right] - \left[\frac{q(k-1)}{r}\right].$$

Let $S_k := \sum_{j=1}^{k} N_j$. Then from the telescoping property of the N_k's, it follows that $S_k = [\frac{qk}{r}]$ and, in particular, that $S_r = q$.

We now prescribe what vertices are to go into each cell. In all cells in a given block of a given row, the red vertices v_h are the same, beginning with $v_1,\ldots,v_r$ in the first cell of the first row and then increasing the subscripts by 1 (cyclically modulo m) each time a new block is reached, moving on from block to block and row to row. The blue vertices begin with $w_1,\ldots,w_s$ in the first cell, $w_{s+1},\ldots,w_{2s}$, in the second, and continue in sequence (modulo n) as new cells are reached. Formally, we have the following entries by block and cell. (It may easily be verified that the two descriptions of entries are indeed the same.)

(S1) The subscripts of the red vertices in each cell of the (i,k) block are, modulo m,

$$(i-1)r+k,\ldots, ir+k-1$$

(S2) The subscripts of the blue vertices in the (i,j) cell are, modulo n,

$$(i-1)qs + (j-1)s+1,\ldots, (i-1)qs + js.$$

We illustrate our construction for a maximum packing of $K_{9,16}$ with copies of $K_{2,3}$. That is ,

$m = 9$, $r = 2$, $p = 4$, and $t = 1$, and
$n = 16$,, $s = 3$, $q = 5$, and $u = 1$.

(Note that $\frac{pu}{s} < \frac{qt}{r}$.) Thus, $\nu = 21$ and $a = 1$. In each of the four full rows, there are two blocks, the first with two cells and the second three. We use the capital letters A to I to denote the nine red vertices and the small letters from a to p for the sixteen blue ones. The complete array is shown in Table 1, and it is easily seen that no edge appears in two of the $K_{2,3}$'s, so $\mu_{2,3}(9,16) = \left[\frac{16}{3}\ [\frac{9}{2}]\right] = 21.$

TABLE 1.

AN OPTIMAL PACKING OF $K_{9,16}$ WITH $K_{2,3}$.

AB abc	AB def	BC ghi	BC jkl	BC mno
CD pab	CD cde	DE fgh	DE ijk	DE lmn
EF opa	EF bcd	FG efg	FG hij	FG klm
GH nop	GH abc	HI def	HI ghi	HI jkl
IA mno				

It is clear that, not only in this example but in general, a red vertex v_x with $r \leq x \leq m$ is adjacent in the subgraphs to cyclically consecutive blue vertices. Since these are at most sq in number, no repetitions can occur. Therefore, if it could be shown that no repetitions occur for v_x when $1 \leq x \leq r-1$, then the construction would establish the packing number, that is, it would verify that $\mu = \nu$. Unfortunately, there are cases in which duplications do occur in this construction; for example, in packing $K_{25,25}$ with $K_{4,4}$'s. In this case, $m = n = 25$, $r = s = 4$, $p = q = 6$ and $t = u = 1$, so $\nu = 37$. In Table 2, we show a portion of this array—

the beginning and the end. Note that edge $B\ell$ occurs in subgraphs in both the first row and the last full row. This can, however, be rectified by deleting the last occurrence of ℓ and advancing subsequent vertices one, so that the last four cells would be

WXYA	XYAB	XYAB	YABC
hijk	mnop	qrst	uvwx

While such a simple deletion will not always suffice, it is always possible to modify the basic construction with a combination of omissions which will yield a packing with ν subgraphs. One way of doing this is the following:

TABLE 2.

AN UNSUCCESSFUL ATTEMPT AT PACKING $K_{25,25}$ WITH $K_{4,4}$

ABCD	BCDE	BCDE	CDEF	DEFG	DEFG
abcd	efgh	ijkl	mnop	qrst	uvwx
UVWX	VWXY	VWXY	WXYA	XYAB	XYAB
uvwx	yabc	defg	hijk	lmno	pqrs
YABC					
tuvw					

Let $\delta := pm - \nu s$. Since a given blue vertex can appear in up to p elements of a packing and since there are s in each element, this represents the "unused" portion of the blue vertices, the "amount of slack" we have to work with. Note that $\delta = pu - s[\frac{pu}{s}]$. Further, let $\gamma := n - qs$, the number of blue vertices which do not appear in a full row. Also, let $\delta = \alpha\gamma + \beta$ with $0 \leq \beta < \gamma$. Our modification of the construction is such that after the first α full rows, we omit γ blue vertices (so $\alpha + 1$ rows are alike in their blue entries), and after the $(\alpha + 1)$th row, we omit β blue vertices. The total number of omissions is thus δ, the amount of slack. Formally, let

$$\varepsilon_i = \begin{cases} (i-1)\gamma & \text{for } i = 1,2,\ldots, \alpha + 1 \\ \delta & \text{for } i = \alpha + 2,\ldots, p + 1 \end{cases}$$

The entries in the array are defined as follows:

(T1) The subscripts of the red vertices in each cell of the (i,k) block are, modulo m,

$$(i-1)r+k,\ldots, ir + k - 1$$

(T2) The subscripts of the blue vertices in the (i,j) cell are, modulo n,

$$\varepsilon_i + (i-1)qs + (j-1)s + 1,\ldots, \varepsilon_i + (i-1)qs + js.$$

We illustrate this final construction in Table 3, in which we pack forty-five $K_{7,7}$'s into $K_{31,80}$. Here,

$m = 31,\ \varepsilon = 7,\ p = 4,\ t = 3;$

$n = 80,\ s = 7,\ q = 11,\ u = 3;$

$a = [\frac{pu}{s}] = 1,\quad \nu = pq + a = 45;$

$N_1 = N_3 = N_5 = 1,\ N_2 = N_4 = N_6 = N_7 = 2$

$\delta = np - \nu s = 5,\ \gamma = n - qs = 3,\ \alpha = 1,\ \beta = 2;$

$\varepsilon_1 = 0,\ \varepsilon_2 = 3,\ \varepsilon_3 = \varepsilon_4 = \varepsilon_5 = 5$

What remains is to show that in this construction no edge appears twice; that is, no red vertex v_x shares two cells with the same blue vertex w_y. We consider separately the cases $1 \leq x \leq r - 1$ and $r \leq x \leq m$. In the latter case, v_x is in consecutive cells (going from one row to the next in the usual way), and these are

TABLE 3

AN OPTIMAL PACKING OF $K_{31,80}$ WITH $K_{7,7}$

1-7 1-7	2-8 8-14	2-8 15-21	3-9 22-28	4-10 29-35	4-10 36-42	5-11 43-49	6-12 50-56	6-12 57-63	7-13 64-70	7-13 71-77
8-14 1-7	9-15 8-14	9-15 15-21	10-16 22-28	11-17 29-35	11-17 36-42	12-18 43-49	13-19 50-56	13-19 57-63	14-20 64-70	14-20 71-77
15-21 80-6	16-22 7-13	16-22 14-20	17-23 21-27	18-24 28-34	18-24 35-41	19-25 42-48	20-26 49-55	20-26 56-62	21-27 63-69	21-27 70-76
22-28 77-3	23-29 4-10	23-29 11-17	24-30 18-24	25-31 25-31	25-31 32-38	26-1 39-45	27-2 46-52	27-2 53-59	28-3 60-66	28-3 67-73
29-4 74-80										

at most q in number. Therefore it shares cells with at most qs blue vertices. Since these are cyclically consecutive except possibly for one jump of up to γ (which equals $n - qs$), there can be no duplications among them.

Now assume that $1 \leq x \leq r - 1$. Then v_x is adjacent to the first sS_x blue vertices at the beginning of the array. The proof will therefore be complete if it is shown that the other blue vertices (if any) to which v_x is adjacent are consecutive and exceed sS_x. To this end, we determine the block in which v_x first reappears (assuming it does). From (T1), it follows that the last red vertex in the $(p-1)$th row is number $pr-1$, so v_x cannot reappear until the pth or $(p+1)$th row. It is readily seen that it will first be in the $(x+t+1)$th block after the $(p-1)$th row.

According to (T2), the last blue entry in the preceding cell is congruent modulo n to $\delta + (p-1)qs + sS_{x+t}$. We now show that this is at least $sS_x + (p-1)n$. For,

$$\delta + (p-1)qs + sS_{x+t} - sS_x - (p-1)n$$

$$= pn - (pq+a)s + (p-1)qs - (p-1)n + s(S_{x+t} - S_x)$$

$$= n - qs + s\left(\left[\frac{(x+t)q}{r}\right] - \left[\frac{xq}{r}\right] - \left[\frac{pu}{s}\right]\right) \geq 0,$$

using the fact that $\frac{pu}{s} \leq \frac{qt}{r}$ and an elementary property of the greatest integer function. We also note that the last blue entry, being in the $(p+1, a)$ cell, is congruent to $\delta + pqs + a = pn$. It therefore follows that if the red vertex v_x is in a cell with a blue vertex w_y in one of the last two rows, then $sS_x + 1 \leq y \leq n$. Hence there can be no duplication of edges anywhere in the array and thus $\mu \geq \nu$. This completes the proof of our main result.

Theorem. The maximum number of graphs in a packing of $K_{m,n}$ with $K_{r,s}$ is

$$\mu_{r,s}(m,n) = \min\left\{\left[\frac{m}{r}\left[\frac{n}{s}\right]\right], \left[\frac{n}{s}\left[\frac{m}{r}\right]\right]\right\} .$$

REFERENCES

1. I. Anderson, A First Course in Combinatorial Mathematics. Clarendon Press, Oxford (1974), MR49#2402.
2. L. W. Beineke, A survey of packing and coverings of graphs. The Many Facets of Graph Theory. (G. Chartrand and S.F. Kapoor, eds.) Springer Lecture Notes in Mathematics 110(1969), 45-53; MR41#1562.
3. L. W. Beineke and R. K. Guy, The coarseness of the complete bipartite graph. Canad. J. Math. 21(1969), 1086-1096; MR41#6727.
4. G. Chartrand, D.P. Geller, and S.T. Hedetniemi, Graphs with forbidden subgraphs. J. Combin. Theory (B) 10(1971), 12-41; MR44#2645.
5. R.K. Guy, A problem of Zarankiewicz. Theory of Graphs (P. Erdös and G. Katona, eds.) Academic Press, New York (1968), 119-150; MR39#4031.
6. C.A. Rogers, Packing and Covering. Cambridge University Press, Cambridge, (1964), MR30#2405.
7. N.J.A. Sloane, The Packing of Spheres, Scientific American, 250(1984), 116-125.

CONNECTIVITY AND SYMMETRY IN GRAPHS

F. Boesch* and R. Tindell

Stevens Institute of Technology
Hoboken, NJ 07030

ABSTRACT. Regular graphs with point- or line-connectivity equal to their degree are of interest in the design of reliable networks. Several classes of graphs exhibiting some symmetry properties are known to achieve these connectivities. We survey the theory of symmetric graphs and its relation to connectivity properties. Several new results related to a generalized connectivity concept called superconnectivity are also presented.

1. INTRODUCTION

The terminology and notation herein follows Harary [8]. A well-known connectivity extremal problem that arises in the design of reliable networks is to determine the minimum value of the number of lines q in an n-connected graph ($\kappa \geq n$) on p-points, for given p and n. It is easily verified that $q \geq \lceil np/2 \rceil$.

There are in fact graphs realizing this lower bound for any p and n. If p or n is even, such graphs must

*Research supported in part under NSF Grant ECS-8100652.

be regular of degree $\delta = \kappa = n$. However, regular graphs need not have $\kappa = \delta$; the most obvious example, the smallest cubic graph with a bridge in Figure 1, has poor "symmetry" properties. The original solution to the extremal problem by Harary [7] used "highly-symmetric" graphs. Our purpose is to survey the main results concerning the connectivity structure of symmetric graphs.

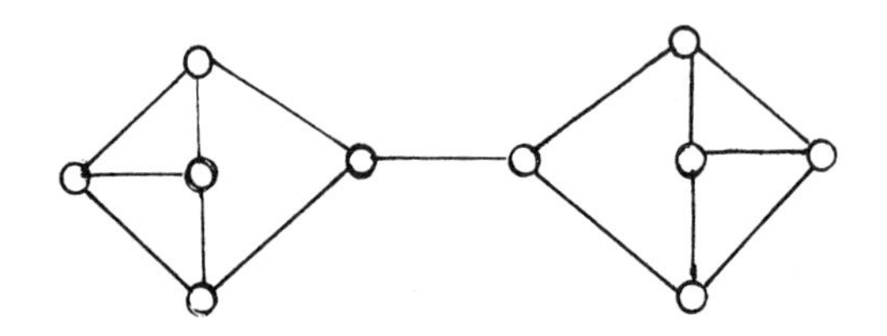

Figure 1. A regular graph with $\kappa < \delta$

2. SYMMETRY

The notions of symmetry that we shall consider are those involving the ability to map arbitrarily points to points or lines to lines by graph automorphisms. Recall that an *automorphism* on G is an adjacency preserving bijection on the point set of G. The set of automorphisms of G forms a group under composition. Two points u and v are *similar points* if there is an automorphism mapping u onto v. Note that point similarity is an equivalence relation. A graph in which no two distinct points are similar is called an *identity graph*. A graph in which all pairs of points are similar is called *point-symmetric*. Clearly a point-symmetric graph is regular. The graph of Figure 2, however, is a regular identity graph; it has the minimum number of points for a regular nontrivial identity graph.

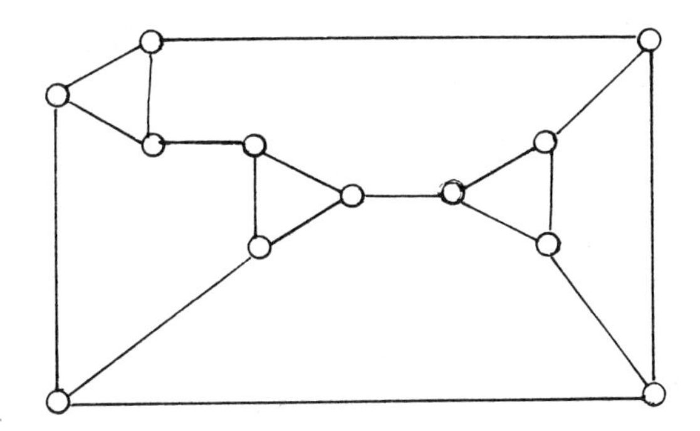

Figure 2. A minimum regular identity graph

Two lines x and y of G are said to be similar lines if there is an automorphism that maps the set of two points of x onto those of y. For a complete discussion of the distinction involved in the alternate approach to defining similar lines via line set bijections, see Behzad, Chartrand, and Lesniak-Foster [2]. A graph is said to be line-symmetric if all pairs of lines are similar. Note that $K_{1,2}$ is line-symmetric but not point-symmetric and $K_2 \times K_3$ is point-symmetric but not line-symmetric. Examples of regular, line-symmetric graphs which are not point-symmetric are more difficult to construct. The minimum number of points in such an example was shown by Folkman [6] to be 20 (see Figure 3).

A stronger notion of line-symmetry, introduced by Tutte [19], is defined as follows. Two lines of G, $x = \{u,x\}$ and $y = \{s,t\}$ are reflexively similar if there are automorphisms ϕ and ψ such that $\phi(u) = s$, $\phi(v) = t$

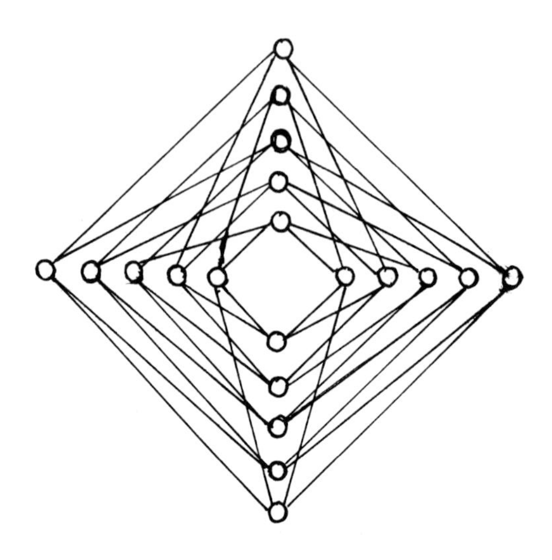

Figure 3. The Folkman graph on 20 points.

and $\psi(u) = t$, $\psi(v) = s$. A graph $G \neq \bar{K}_p$ in which any pair of lines are reflexively similar is said to be 1-transitive. For example, the graphs of the Platonic solids (see Ore [14, p9] are all 1-transitive, and so is the Petersen graph. Clearly, 1-transitivity implies line-symmetry; the converse is false by $K_{1,2}$. Theorem A, whose proof is immediate, gives a characterization of those line-symmetric graphs which are 1-transistive.

Theorem A. A line-symmetric grpah G is 1-transitive if and only if there is a line x reflexively similar to itself.

An obvious connection between 1-transitivity and point-symmetry is that a 1-transitive graph without isolated points is point-symmetric. Notice that a graph which is both point- and line-symmetric is not apriori 1-transitive. However, Tutte [19] proved the following statement.

Theorem B. (Tutte) Every point- and line-symmetric graph of odd degree is 1-transitive.

For the case of even degree, the question was resolved by Bouwer [5] who constructed point- and line-symmetric graphs of even degree that are not 1-transitive; his smallest example has 54 points. The smallest possible graph is not yet determined.

When the number of points is prime, Turner [18] showed that point-symmetric graphs belong to a very special class of graphs. These graphs, known as circulants are defined as follows. To this end we assume the points of a graph are (labelled) $0, 1, \ldots, p-1$. The <u>circulant graph</u> $C_p < a_1, a_2, \ldots, a_k >$ or briefly $C_p < a_i >$, where $0 < a_1 < a_2 \ldots < a_k < (p+1)/2$, has $i \pm a_1$, $i \pm a_2, \ldots, i \pm a_k \pmod p$ adjacent to each point i. The sequence $< a_i >$ is called the <u>jump sequence</u> and the a_i are called the <u>jumps</u>. Notice that our definition precludes jumps of size greater than p/2 as such jumps would produce the same result as a jump of size less than p/2. Also note that if $a_k \neq p/2$ then the circulant is always regular of degree 2k. When p is even we have allowed $a_k = p/2$ (called a <u>diagonal jump</u>), and when $a_k = p/2$ the circulant has degree 2k-1.

The Harary graphs [7] mentioned in the introduction are the circulants having $a_i = i$. The fact that circulants are point-symmetric may be seen by considering the <u>rotation maps</u>, $i \to (i + b) \bmod p$, for each fixed b. The map $j \to (a_i - j) \bmod p$ is an automorphism of $C_p < a_1, \ldots, a_k >$ that shows that the line $\{0, a_1\}$ is reflexively similar to itself. Thus every line-symmetric circulant is 1-transitive. Very few circulants are in fact line-symmetric. For more details on circulants see Boesch and Tindell [3], which contains many connectivity properties of this class of graphs. Here we mention one such property which was conjectured by the authors at this conference, namely that all connected circulants are hamiltonian. Notice that this conjecture is a variant of the more general and still unproven Lovász conjecture.

3. CONNECTIVITY

The extremal problem given in our introduction was solved by Harary [7], who noted that $C_p < 1,2,\dots,k >$ has $\kappa = \delta$. Now it is known that this property does not extend to all point-symmetric graphs. However, Watkins [20],[21] and independently Mader [10-13] have developed some deep structural properties of point-symmetric graphs having $\kappa < \delta$. We summarize some of these important results here. To this end, consider the following definitions. An <u>atomic part</u> part of G is a smallest order (number of points) component of G-S over all point-disconnecting sets S of size κ. The <u>atomic number</u> $\alpha(G)$ of a graph G is the number of points in an atomic part.

Theorem C. (Watkins and Mader)

(A) For a connected point-symmetric graph G with p points:

(i) $\kappa = \delta$ if and only if $\alpha = 1$,

(ii) $\alpha(G)$ divides p,

(iii) $\delta = \lambda \geq \kappa > 2\delta/3$,

(iv) If G does not contain K_4 then $\kappa = \delta$.

(B) For a connected line-symmetric graph, κ is equal to the minimum degree δ.

The circulant $C_9 < 1,2,3 >$ can be used to verify that neither of the conditions (A)(iv) nor (B) are necessary for the $\kappa = \delta$ property. In the case of circulants, however, one can give explicit, necessary and sufficient conditions for $\kappa = \delta$. These conditions, which are derived in [3], are obtained from Theorem C and the following result of Tindell [16].

Theorem D. (Tindell) Let G be a p-point connected circulant with atomic number $\alpha(G)$ and $p = m\alpha$. If A is an atomic part of G then $i \in A(G) \rightarrow (i + m) \bmod p \in A$.

The next theorem is from Boesch and Tindell [3].

Theorem E. The circulants $C_p < a_i >$, $1 \le i \le k$, satisfies $\kappa < \delta$ if and only if for some proper divisor m of p, the number r_m of distinct positive residues modulo m of the numbers $a_1, \ldots, a_k, p - a_k, \ldots, p - a_1$ is less than the minimum of $m - 1$ and $\delta m/p$.

As an example of properties given in Theorem E, we consider $C_8 < 1,3,4 >$. Clearly m is either 2 or 4. For $m = 4$, the sequence 1,3,4,5,7 has exactly $r_m = 2$ distinct positive residues mod 4; two distinct elements corresponding to these two distinct residues are the jumps 1 and 3. Now if $\alpha(G)$ is to equal p/m for $m = 4$ then by Theorem D an atomic part would be defined by the point set $A = \{0,4\}$. Consider now the determination of the deleted neighborhood of A, called N(A) and defined as the set of all points in $G-A$ that are adjacent to at least one point in A. Clearly $i \in A$ implies $(i+a_j) \bmod p \in N(A)$ if and only if $a_j \not\equiv 0 \pmod{m}$. Now let J be a set of r_m jumps having distinct positive residues modulo m; in this example one could choose $J = \{1,3\}$. Then clearly each point $(i+a_j) \bmod p$, with $i \in A$ and $a_j \in J$, is in $N(A)$. Also the sets of points in $N(A)$ obtained from points i, $j \in A$ by adding the jumps in J are pairwise-disjoint. Moreover these sets cover $N(A)$. Hence $|N(A)| = |A|\, r_m$; in our example, $N(A) = \{0+1,\ 0+3\} \cup \{4+1,\ 4+3\}$.

Now an atomic part of a p-point circulant will have as points all multiples of some divisor m of p. The above argument shows that the neighborhood of any such set will have $r_m \cdot a$ elements, where $a = p/m$. Observe that for any nonempty set A, $N(A)$ disconnects the graph if and only if there is at least one point not in $A \cup N(A)$. Moreover, we will have $k < \delta$ if and only if for some such set $|N(A)| < \delta$. These last two conditions translate directly into the conditions of the theorem. By considering all the possible divisors of p, one obtains the actual connectivity. In this example $\delta = 5$ but $\kappa = 4$, corresponding to $A = \{0,4\}$ and point disconnecting set $N(A) = \{1,3,5,7\}$.

In addition to the circulants characterized in Theorem E there are some other classes of graphs which satisfy either $\kappa = \delta$, $\lambda = \delta$ or both. For instance, the regular, complete, n-partite graphs have $\kappa = \lambda = \delta$. In fact, they are circulants and can most easily be defined as $\overline{C_{nb} \langle n, 2n, \ldots, (b-1)n \rangle}$; note that if G is a circulant then its complement $\bar{G}$ is also a circulant. Another class of graphs called "crowns" are defined as follows. Let G be a p-point bipartite graph (p even) having points $0,2,\ldots,p-2$ in one part and points $1,2,\ldots,p-1$ in the other part; crowns are defined by connecting each even number i to the k odd numbers $(i+1, i+3, \ldots, i+2k-1) \bmod (p-1)$ for any positive integer $k (k \leq p/2)$. Generally, crowns are not circulants; however, they have the property that $\kappa = \lambda = \delta$. Finally we mention that clearly any line-disjoint union of hamiltonian cycles has $\lambda = \delta$; for example the graph $K_2 \times K_4$ consists of 2 line-disjoint hamilton cycles. Using a decomposition of the complete graph (p odd) into hamilton cycles, we obtain a class of $\lambda = \delta$ graphs for any even δ; one such possible decomposition is given in Harary [8].

4. STRUCTURE OF DISCONNECTING SETS

We now turn to a study of the nature of the disconnecting sets of graphs, particularly the regular graphs of maximum connectivity.

In a regular graph G with $\kappa = \delta$ the neighborhood of each point will be a minimum size disconnecting set of points for G; if these are the only minimum size disconnecting sets for G we shall refer to G as a super-κ graph. Similarly, a regular graph with $\lambda = \delta$ and such that each minimum size disconnecting set of lines isolates a point is called a super-λ graph. For the relevance of such graphs to the construction of reliable networks, see [1]. We first examine the super-λ property relative to the classes of graphs introduced above.

First we consider the class of circulants. In [1] it was shown that a wide class of circulants are super-λ.

Theorem F. (Bauer et al.) If $k \geq 2$, then $C_p\langle 1, a_2, \ldots, a_k\rangle$ is super-λ.

It should be noted that the above theorem does not characterize the super-λ circulants, as may be seen by considering $C_{10}\langle 3,5\rangle$, which is super-λ, and is isomorphic to $C_{10}\langle 1,5\rangle$. A complete classification of super-λ circulants is derived in Boesch and Wang [4].

Theorem G. (Boesch and Wang) The only connected circulants which are not super-λ are the cycles and the graphs $C_{2n}\langle 2,4,\ldots,n-1,n\rangle$ with $n \geq 3$ an odd integer.

It should be noted that by the above, not all connected point-symmetric graphs are super-λ. In addition to the non super-λ circulants, there are many point-symmetric graphs which are not super-λ. One such example is the non-circulant $K_2 \times K_4$. The story is different with line-symmetric graphs, cf. Tindell [17].

Theorem H. (Tindell) Every connected line-symmetric graph is super-λ.

In the same paper in which the above is proved, a characterization of connected symmetric graphs which are not super-λ is given. We now describe this result. Given an r-regular graph G, define the <u>s-fold expansion</u> $\xi_s(G)$ of G, the rs-regular graph to be obtained as follows:

1. For each vertex u of G choose a complete graph H_u on $r\cdot s$ points in such a way that H_u is disjoint from H_v if $u \neq v$.
2. Add $q\cdot s$ independent lines to the union of the H_u's in such a way that there are exactly s lines connecting H_u to H_v if u and v are adjacent in G. (Note that by counting we see that there are

no lines from H_u to H_v if u and v are not adjacent in G.)

By the way of example $\xi_3(K_2)$ and $\xi_2(C_4)$ are given in Figure 4. The next theorem is also proved in [17].

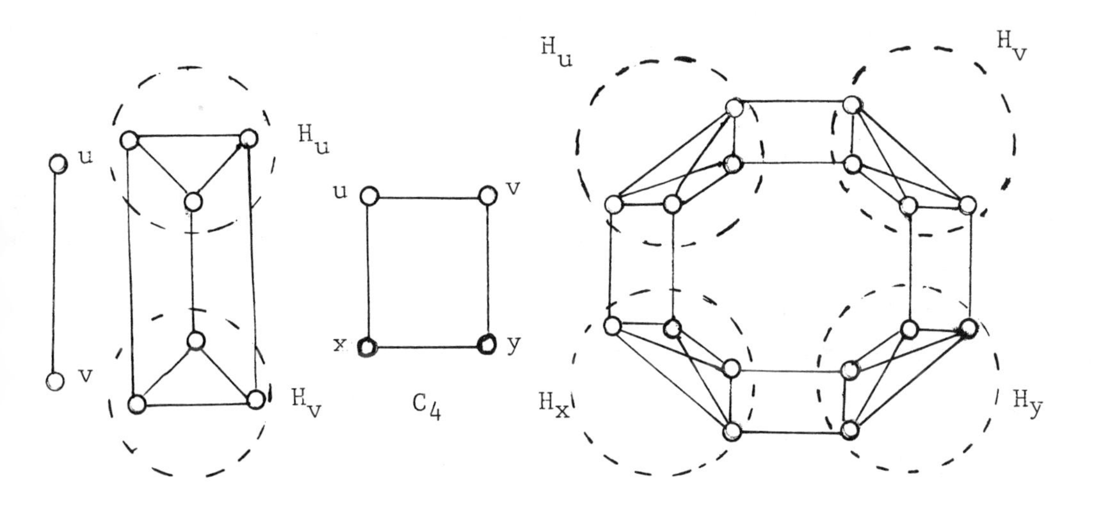

Figure 4.

Theorem I. (Tindell) A connected graph $G \neq C_p$ is point symmetric but not super-λ if and only if $G = \xi_s(H)$ for some connected 1-transitive graph H and $s > 1$.

We now turn to the super-κ graphs, the study of which turns out to be considerably more difficult than the above. A few isolated results are known. For example the regular, complete, n-partite graphs are super-κ. Some other special cases are given by Smith [15]. However, general theorems of the type obtained above are not known for super-κ graphs. By way of example, we note that the circulant $C_{12} < 1,5 >$ is 1-transitive and is not super-κ: the points 2,5,8,11 disconnect $C_{12}\langle 1,5\rangle$ without isolating a point. Thus no simple super-κ version of the above theorem is possible.

The analogue of Theorem G (which is to say finding a characterization of super-κ circulants) appears to be very difficult. For a discussion of the problem and a statement of some partial results due to Wang, see [3].

5. CONCLUSIONS

We have considered here the extremal problem of determining regular graphs of a given degree δ and a given number of points p such that κ or λ equals δ. In addition, we considered the case where such graphs are also super-λ or super-κ. It was seen that connectedness and point-symmetry alone are not sufficient for any of these properties excepting $\lambda = \delta$; a structural result characterizes the non-super-λ case. Explicit necessary and sufficient conditions for the special class of point-symmetric graphs, known as circulants, to have $\kappa = \delta$ were given. In addition, the necessary and sufficient conditions for circulants to be super-λ were stated. Connected, line-symmetric graphs are super-λ, and have $\kappa = \lambda = \delta$. The super-$\kappa$ case appears to be the most difficult to describe as the symmetry property 1-transitivity (which is stronger than the combination of point- and line- symmetry) does not suffice for super-κ.

In view of the above observations, we might question the relevance of the symmetry concepts to the connectivity extremal problem. For example, the Folkman graph (shown in Figure 3) has $\kappa = \delta$, but it is not point-symmetric. Perhaps even more surprising is the fact that the regular identity graph of Figure 2 has $\kappa = \lambda = \delta$; however, no two distinct points or lines of this graph are similar. Although the identity graph of Figure 2 is neither super-κ nor super-λ , we suspect that there exist super-κ , super-λ identity graphs. This conjecture is based in part on the results of Izbicki [9] who showed that there are regular graphs with many prescribed properties that correspond to a prescribed automorphism group.

REFERENCES

1. D. Bauer, F. Boesch, C. Suffel, and R. Tindell, Connectivity extremal problems and the design of reliable probabilistic networks. The Theory and Applications of Graphs (G. Chartrand, Y. Alavi, D. Goldsmith, L. Lesniak-Foster, and D. Lick, Eds.) Wiley, New York, (1981), 45-54.
2. M. Behzad, G. Chartrand, and L. Lesniak-Foster, Graph and Digraphs, Wadsworth, Belmont, (1979).
3. F. Boesch and R. Tindell, Circulants and their connectivities. J. Graph Theory (to appear).
4. F. Boesch and J. Wang, Super line-connectivity properties of circulant graphs. (submitted).
5. I.Z. Bouwer, Vertex and edge transitive, but not 1-transitive graphs. Canad. Math. Bull. 13 (1970), 231-237.
6. J. Folkman, Regular line-symmetric graphs. J. Combinatorial Theory 3 (1967), 215-232.

7. F. Harary, The maximum connectivity of a graph. Proc. Nat. Acad. of Sci. USA 48 (1962), 1142-1146.
8. F. Harary, Graph Theory, Addison-Wesley, Reading, (1969).
9. H. Izbicki, Regular Graphen beliegigen Grades mit vorgegebenen Eigenschaften. Monatsh. Math. 64 (1960), 15-21.
10. W. Mader, Minimale n-fach zusammenhangende Graphen mit maximaler Kantenzahl. J. Reine Angew. Math. 249 (1971), 201-207.
11. W. Mader, Über den Zusammenhang symmetrischer Graphen. Arch. Math. (Basel) 21 (1970), 331-336.
12. W. Mader, Eine Eigenschaft der Atome endlicher Graphen. Arch. Math. (Basel) 22 (1971), 333-336.
13. W. Mader, Minimale n-fach kantenzusammenhängende Graphen. Math. Ann. 191 (1971), 21-28.
14. O. Ore, Theory of Graphs, American Math. Society, Providence, (1962).
15. D. Smith, Graphs with the smallest number of minimum cut sets, Networks (to appear).
16. R. Tindell, The connectivities of a graph and its complement. Ann. Discrete Math. 13 (1982), 191-202.
17. R. Tindell, Edge connectivity properties of symmetric graphs. (submitted).
18. J. Turner, Point-symmetric graphs with a prime number of points. J. Combinatorial Theory 3 (1967), 136-145.
19. W.T. Tutte, Connectivity in Graphs, Univ. of Toronto Press, London, (1966).
20. M.E. Watkins, Connectivity of transitive graphs. J. Combinatorial Theory 8 (1970), 23-29.

21. M.E. Watkins, Some classes of hypoconnected vertex-transitive graphs. Recent Progress in Combinatorics (Proc. Third Waterloo Conf. on Combinatorics, W.T. Tutte and C.St.J.A. Nash-Williams, Eds.) Academic Press, New York, (1969), 323-328.

THE PETERSEN GRAPH

Gary Chartrand

Department of Mathematics
Western Michigan University
Kalamazoo, MI 49008

Robin J. Wilson

Mathematical Institute
Oxford OX1 3LB
England

1. ORIGINS AND DEPICTIONS

It is remarkable how very often one particular graph is encountered throughout graph theory. The graph to which we refer is the famous <u>Petersen graph</u> which is usually drawn, as in Figure 1, with an outer 5-cycle, an inner

5-cycle, and five 'spokes' joining them. The object of this expository article is to describe the main properties of the Petersen graph, thereby attempting to explain its fascination and importance. Let P denote the Petersen graph.

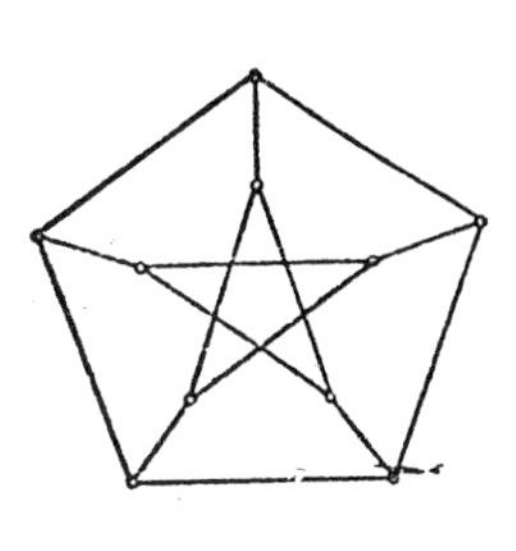

Fig. 1

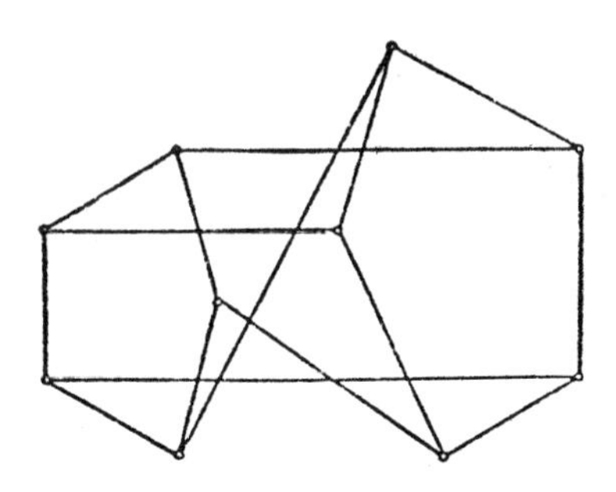

Fig. 2

The origins of the Petersen graph in the literature of graph theory are somewhat obscure. It certainly takes its name after Julius Petersen (1839-1910), a Danish mathematician who became a professor of mathematics at the University of Copenhagen and who wrote several textbooks, one of which [29] achieved international recognition. Although his research interests were wide, including algebra, analysis, number theory and mechanics, his most important work was in geometry. In particular, he spent several years investigating the decomposition of regular graphs into cycles and nonadjacent edges (Section 4), and in a short note [31] in 1898 he introduced the Petersen graph in the form shown in Figure 2. (The drawing in Figure 1 apparently dates from a manuscript of C.S. Peirce in 1903, which was recently published in [14, p. 434].)

However, Petersen was not the first to display P as this graph appeared in an 1886 article [22] by A.B. Kempe, who is known primarily for his erroneous proof of the Four Color Theorem. (It may be worth noting that

Petersen was one of the few reputable mathematicians to have stated publicly [32] that he believed the Four Color Theorem to be false.) In Kempe's paper P appears in the form shown in Figure 3 in connection with 'the theorem that if two coplanar triangles are coaxial, they are also copolar'. This is Desargues' theorem on triangles in perspective (Figure 4), and P is obtained by joining those pairs of points (such as a and d) that do not lie on the same line.

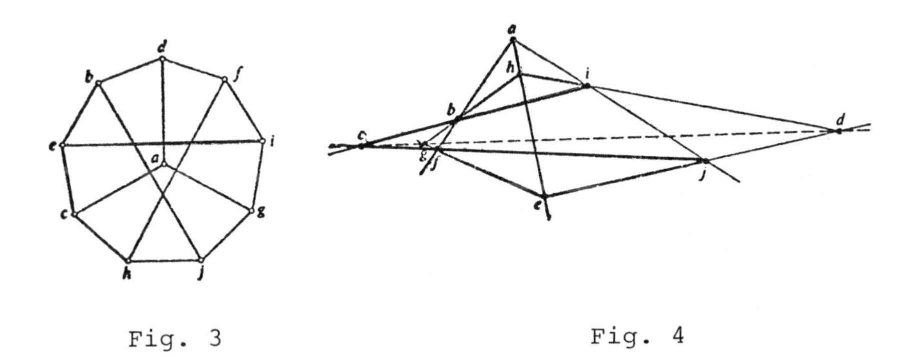

Fig. 3

Fig. 4

The Petersen graph also arises in other contexts. For example, it can be obtained by identifying opposite vertices of a dodecahedron (Figure 5). Alternatively, as was observed by Kowalewski [23] in 1917, P may be obtained by taking the ten 2-subsets of $\{1,2,3,4,5\}$ and joining those pairs of subsets that are disjoint (Figure 6). This last description suggests that P has a high degree of symmetry, and we can also illustrate this by re-drawing the graph as in Figure 7. In this drawing we can take any vertex as the central vertex, and its 'neighbors' and 'non-neighbors' can then be arranged accordingly. Yet another drawing of P appears in Figure 8. This drawing, due to Erdös, Harary and Tutte [15], has the property that every edge is represented by a

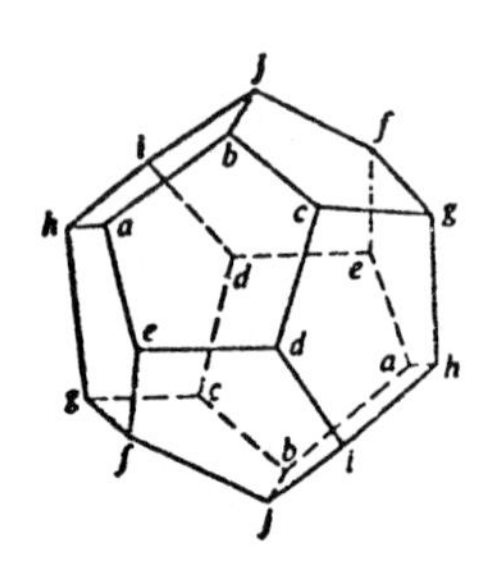

Fig. 5

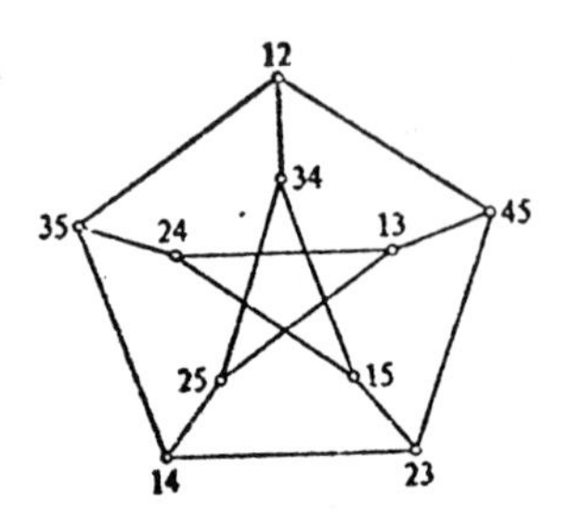

Fig. 6

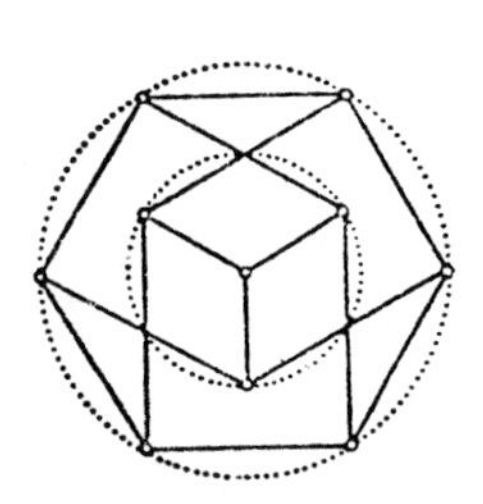

Fig. 7

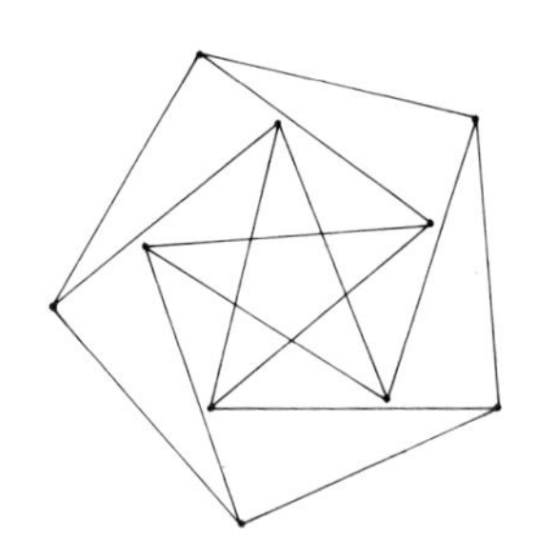

Fig. 8

straight line section of unit length. Finally, P also appears in the study of chemical reactions as a 'regular halved combination graph' [3].

The graph-theoretical terminology used in this paper is standard. All terms may be found in, for example, [5], [19] or [41].

2. SOME CHARACTERIZATIONS

It is clear that P is a 3-regular graph with 10 vertices, 15 edges, and girth 5. What is not immediately clear is that P is the smallest 3-regular graph with girth 5. If we define an <u>r-cage</u> to be a 3-regular graph of smallest order with girth r, then it is easy to see that the complete graph K_4 is the only 3-cage, and that the complete bipartite graph $K_{3,3}$ is the only 4-cage. The above characterization of the Petersen graph can now be presented.

Theorem 1. The Petersen graph is the only 5-cage.

Proof. Let G be a 5-cage, let v be an arbitrary vertex of G, and let w_1, w_2 and w_3 be the three vertices adjacent to v. Since G has girth 5, no two of w_1, w_2 and w_3 can be adjacent. For i = 1,2,3, let x_i and y_i be the vertices (other than v) that are adjacent to w_i. Since G has girth 5, these vertices are all distinct, and each is adjacent to only one of the vertices w_i. The situation is now as depicted in Figure 9.

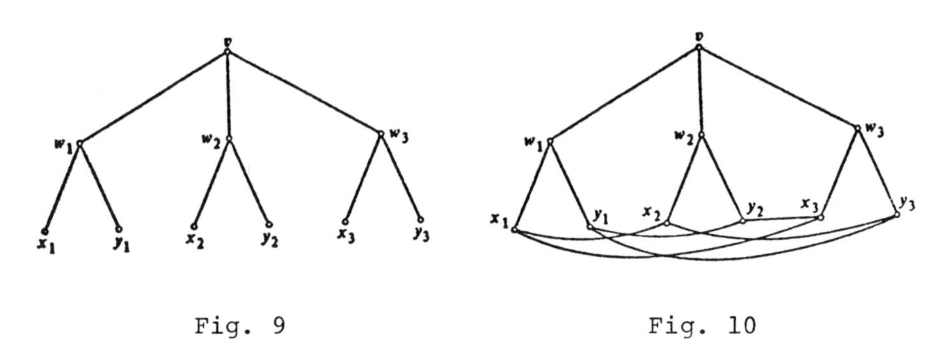

Fig. 9 Fig. 10

Since P is a 3-regular graph with 10 vertices and girth 5, and since G has at least 10 vertices, it follows that the Petersen graph is a 5-cage and that G has exactly 10 vertices. In order to show that G is in fact P (that is, that there are no other 5-cages), we observe that, since G is 3-regular with girth 5, the vertex x_1 must be adjacent to one of x_2 and y_2 and to one of x_3 and y_3. Assuming, without loss of generality, that x_1 is adjacent to x_2 and x_3, we see that y_1 must be adjacent to

y_2 and y_3, and that G must also contain the edges x_2y_3 and y_2x_3 but no others. The resulting graph, shown in Figure 10, is easily seen to be isomorphic to P, so that the Petersen graph is the only 5-cage. □

For more information on r-cages, see [42].

Related to this is the study of Moore graphs. If G is a k-regular graph with diameter d, then it is easy to see that the order of G is at most

$$1 + k + k(k-1) + k(k-1)^2 + \ldots + k(k-1)^{d-1},$$

since any given vertex must have k neighbors, each of which has $k - 1$ additional neighbors, and so on. A k-regular graph with diameter d whose order is equal to this expression is called a <u>Moore graph</u>. It has been proved that if G is a Moore graph of diameter 2, so that G has order $k^2 + 1$, then the only possible values of k are 2, 3, 7 and (perhaps) 57. In the case $k = 3$, we can use the above method of proof to show that there is only one Moore graph--namely, the Petersen graph. For the other cases, see [8, Chapter 23].

We conclude this section by stating two further results. The first uses the fact that any two non-adjacent vertices of P are mutually adjacent to just one other vertex. The second result is due to Murty [28].

Theorem 2. Apart from the complete graph K_4, the Petersen graph is the only 3-regular graph in which any two non-adjacent vertices are mutually adjacent to just one other vertex.

Theorem 3. The Petersen graph is the smallest graph with the property that, given any three distinct vertices u, v and w, there is a fourth vertex adjacent to u but not to v or w.

3. SYMMETRY

It is clear from Figures 1, 3 and 7 that P has a high degree of symmetry. This can be made precise in a variety of ways. In this section we consider the symmetry of P from three different points of view.

3.1. Transitivity

We notice first that P is vertex-transitive, that is, it has automorphisms mapping any given vertex to any other. To see this, consider P as drawn in Figure 6, where the vertices are the 2-subsets of $\{1,2,3,4,5\}$. It is clear that any permutation of $\{1,2,3,4,5\}$ gives rise to an automorphism of the graph, and we can therefore obtain an automorphism taking any given vertex $\alpha\beta$ to any other vertex $\gamma\delta$ by choosing a permutation that maps α to γ and β to δ. In fact we can say more than this, since it can be shown that there are automorphisms taking any path of length 3 to any other. For example, the path abcc' in Figure 11 is mapped onto the path a'c'cd by the permutation (145)(23).

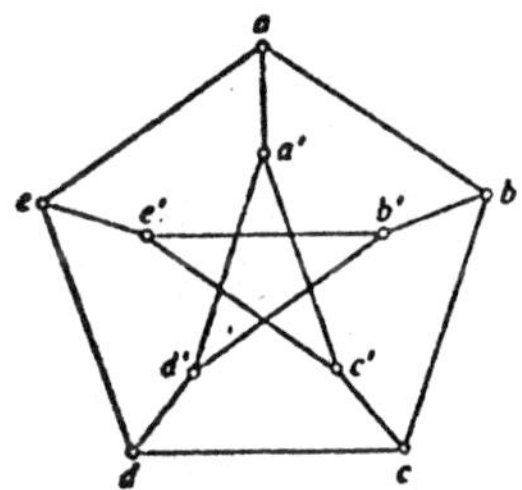

Fig. 11

It is also easy to see that P has the property that there are automorphisms taking any given pair of adjacent vertices to any other such pair, and there are automorphisms taking any given pair of nonadjacent vertices to any other such pair. More generally, P is distance-transitive in the sense that, whenever the distance from v to w is the same as that from v' to w', there is an automorphism taking v to v' and w to w'. Graphs of this kind can be drawn as in Figure 7, where any vertex v may be taken as the central vertex, and the vertices at various distances from v are arranged in circles around it. It can be shown (see [8, Chapter 21]) that there are only twelve 3-regular distance-transitive graphs.

We conclude this discussion of transitivity by remarking that, unlike many vertex-transitive graphs, P is not a Cayley graph. In the connection, a Cayley graph is a graph obtained from a finite group Γ with generating set $\Omega = \{g_i\}$ (where $1 \notin \Omega$, and g_i^{-1} is also in Ω for each i) by taking the elements of Γ as vertices, and joining two vertices v and w whenever $v = wg_i$ for some i. Although every Cayley graph is vertex-transitive, the converse statement does not hold in general, and the simplest counter-example is the Petersen graph.

3.2 The Automorphism Group

Since any permutation of {1,2,3,4,5} gives rise to an automorphism of P, it follows that P has at least 120 automorphisms. That there are no other automorphisms can be proved as follows:

Theorem 4. The automorphism group of the Petersen graph is isomorphic to the symmetric group S_5.

Proof. The proof rests on three facts:

(i) The automorphism group of a graph G is isomorphic to that of its complement $\bar{G}$ (since automorphisms preserve non-adjacencies as well as adjacencies).

(ii) With four small-order exceptions, the automorphism group of a connected graph G is isomorphic to that of its line graph L(G). (This follows from a result of Whitney [40].)

(iii) The Petersen graph is the complement of the line graph of K_5. (Figure 12 shows the line graph of K_5, which turns out to be the graph obtained from the Desargues configuration (Figure 4) by joining those pairs of points that lie on a common line.)

Since the automorphism group of K_5 is isomorphic to the symmetric group S_5, the result follows. □

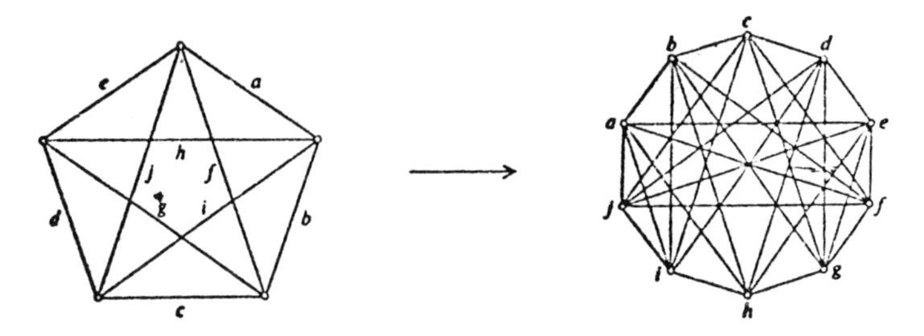

Fig. 12

3.3 Strongly Regular Graphs and Eigenvalues

A strongly regular graph with parameters (n, k, λ, μ) is a k-regular graph of order n in which the number of vertices mutually adjacent to any pair of adjacent vertices is λ, and the number of vertices mutually adjacent to any pair of non-adjacent vertices is μ. Thus the Petersen graph is a strongly regular graph with parameters

(10, 3, 0, 1). It can be shown (see [12]) that a regular connected graph G is strongly regular if and only if its adjacency matrix has just three distinct eigenvalues. If G is k-regular, then these eigenvalues are k, r and s, which are related to λ and μ by the equations

$$\lambda = k + r + s + rs \text{ and } \mu = k + rs.$$

The multiplicities of these eigenvalues are 1 (for the eigenvalue k) and

$$\tfrac{1}{2}\left\{n - 1 \pm \frac{(n-1)(\mu-\lambda) - 2k}{\{(\mu-\lambda)^2 + 4(k-\mu)\}^{\frac{1}{2}}}\right\}.$$

For the Petersen graph, the eigenvalues are 3 (with multiplicity 1), 1 (with multiplicity 5) and -2 (with multiplicity 4), and it can be shown that there are no other graphs with these eigenvalues (see [34]). But if G is a k-regular connected graph with eigenvalues k, $\lambda_1, \ldots, \lambda_{n-1}$, then the number of spanning trees of G is

$$\frac{1}{n}(k - \lambda_1)(k - \lambda_2) \ldots (k - \lambda_{n-1})$$

(see [34]). It follows from this that P has exactly 2000 spanning trees.

4. FACTORIZATION AND COLORING

In the 1890's Petersen studied factorizations of regular graphs. In particular he wrote an important paper [30] proving that every 3-regular bridgeless graph can be factored into a 2-factor and a 1-factor. (He actually proved a somewhat stronger result--that every 3-regular graph with at most two bridges can be so factored.) In an earlier paper in 1880, P. G. Tait [36] wrote that he had previously shown that every 3-regular graph is 1-factorable, but that this result was "not true without limitation." Tait's remarks were at best confusing, and in 1888 Petersen interpreted Tait's 'theorem' as meaning that every 3-regular bridgeless graph is 1-factorable. This result, if true, would have been stronger than Petersen's theorem, but Petersen showed it to be false by producing a 3-regular bridgeless graph that is not 1-factorable--namely, the Petersen graph, as shown in Figure 2.

It is clear that P can be split into a 2-factor and a 1-factor. It is less clear, however, that P provides a counter-example to Tait's theorem. We present two proofs of this fact:

Theorem 5. The Petersen graph is not 1-factorable.

First proof. It is clear that if S is the set of 'spokes' in Figure 1, then S is a 1-factor. If P were 1-factorable, then it could be factored into three 1-factors at least one of which (T, say) would contain at least two spokes. Deleting two such spokes and their incident vertices from P gives the graph shown in Fig. 13, where the edges belonging to S are so labeled. The only 1-factor of this subgraph consists of the three edges in

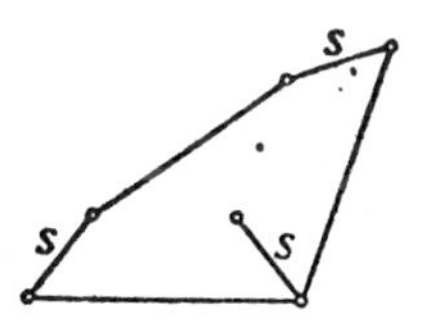

Fig. 13

S, and so T must be identified with S. But this implies that the two 5-cycles can be factored into two 1-factors, which is easily seen to be impossible. □

The second proof we give of this theorem involves coloring the edges of P. If G is a 3-regular graph that is 1-factorable, then its edges can be colored with three colors--simply by assigning a different color to each 1-factor. We prove that P is not 1-factorable by showing that its edges cannot be colored with just three colors.

<u>Second proof</u>. Suppose that the edges of P can be colored with the three colors α, β and γ. Without loss of generality, we may assume that the colors around the outer 5-cycle are α, β, α, β, γ (see Figure 14), since any other permissible assignment of colors can be reduced to this one by a suitable renaming of the colors. Since the three colors appearing at each vertex must be different, the edges bb' and cc' must each be colored γ and the edge ee' must be colored α. It follows that the edges b'e' and c'e' must both be colored β, which is not permissible (see Figure 15). This contradiction yields the result. □

The same argument can be used to show that, even if we remove a vertex (and its incident edges) from P, the

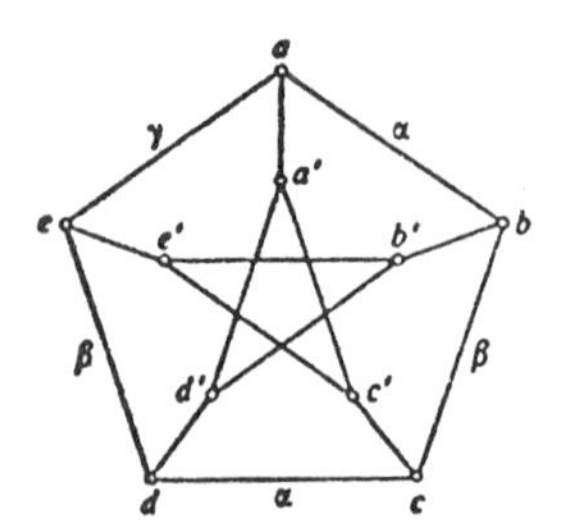

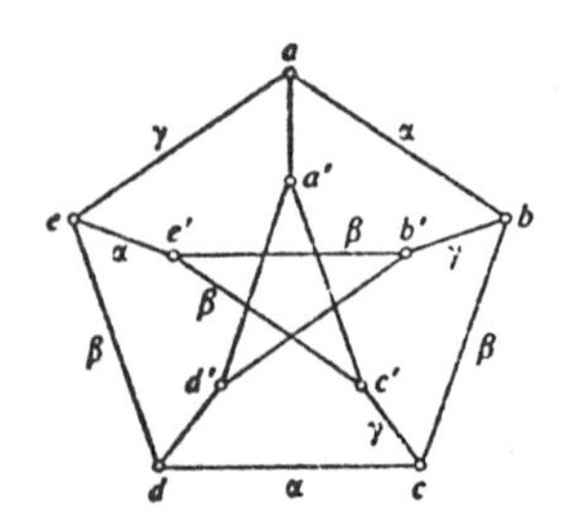

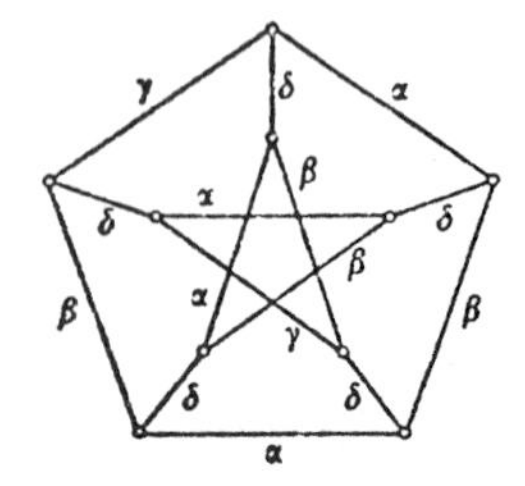

Figs. 14, 15, 16

resulting graph cannot be edge-colored with just three colors. However, both this graph and the Petersen graph itself can be edge-colored with four colors, and a 4-coloring of the edges of P is given in Figure 16. Thus the chromatic index of P is 4.

We can also ask for the chromatic number of the Petersen graph. It is clear that at least three colors are needed, since the outer 5-cycle alone cannot be colored with only two colors. On the other hand, Brooks' theorem [11] tells us that every 3-regular connected graph other than K_4 can be vertex-colored with only three colors. Such a 3-coloring is given in Figure 17.

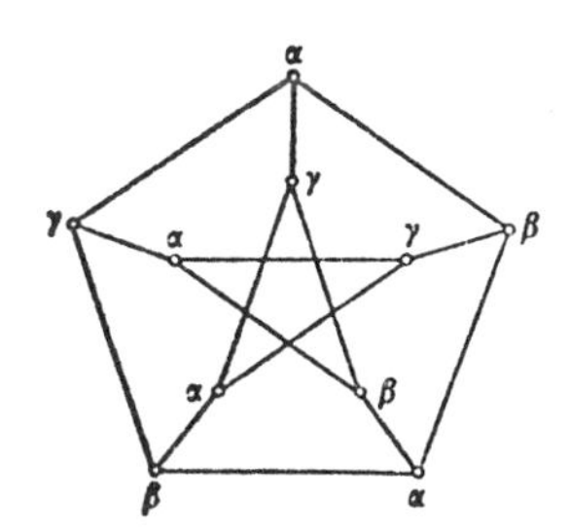

Fig. 17

We conclude this section by observing that, although P is not 1-factorable, it can be factored into three copies of any of the ten graphs shown in Figure 18; this result is due to Alspach and Ruiz (personal communication).

Alspach and Ruiz also showed that there is no factorization of K_{10} into three copies of the Petersen graph; they proved this by exploiting the automorphism group and the fact that the neighbors of any vertex cannot be adjacent. An alternative proof involving eigenvalues has recently been given by Schwenk [33].

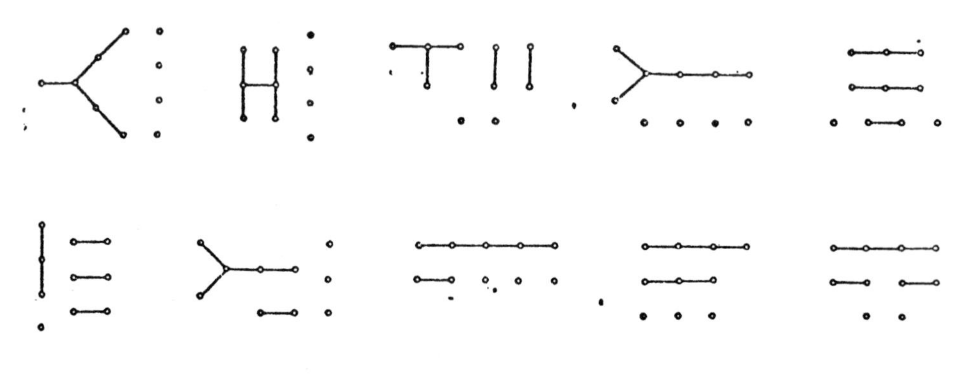

Fig. 18

Theorem 6. K_{10} cannot be factored into three copies of the Petersen graph.

Proof. Suppose that such a factorization exists, and that the three Petersen graphs have adjacency matrices A_1, A_2 and A_3. Then

$$A_1 + A_2 + A_3 = J - I ; \tag{1}$$

where J is the all-1 matrix. Now, for each A_i, $\underline{v} = (1,1,\ldots,1)^T$ is an eigenvector corresponding to the eigenvalue $\lambda = 3$, and so the eigenspaces corresponding to $\lambda = 1$ and $\lambda = -2$ lie in the 9-dimensional orthogonal complement S of $\underline{v}$. For $\lambda = 1$, the eigenspaces of A_1 and A_2 are both 5-dimensional subspaces of S, and so there exists a vector $\underline{w}$ such that $A_1\underline{w} = A_2\underline{w} = \underline{w}$. By (1), $A_1\underline{w} + A_2\underline{w} + A_3\underline{w} = J\underline{w} - I\underline{w}$, and $J\underline{w} = 0$, since $\underline{v}$ is orthogonal to $\underline{w}$. So $A_3\underline{w} = -3\underline{w}$, contradicting the fact that the only eigenvalues of A_3 are 3, 1 and -2. □

5. HAMILTONIAN PROPERTIES

In this section we consider paths and cycles in P,

with particular reference to hamiltonian paths and cycles. It is easy to verify that the Petersen graph has cycles of lengths 5, 6, 8 and 9, but no cycles of length 7. In addition P has no cycles of length 10; we present two proofs of this fact--a direct one, and a proof involving edge-colorings.

Theorem 7. The Petersen graph is not hamiltonian.

First proof. If P were hamiltonian, then it would contain a hamiltonian cycle C. Since this cycle cannot contain every edge of the outer 5-cycle (see Figure 1), we can assume without loss of generality that it does not contain the edge cd, and hence contains the edges bc, cc', de and dd'. The cycle C must also contain one of the edges ab and ea—let us assume without loss of generality that it contains the edge ea. The situation is now as in Figure 19. Since C cannot contain the edge ee', it must contain both b'e' and c'e'. This implies that C does not contain the edge a'c', so that it must contain both aa' and a'd'. It follows that C contains the cycle aa'd'dea, which is impossible. □

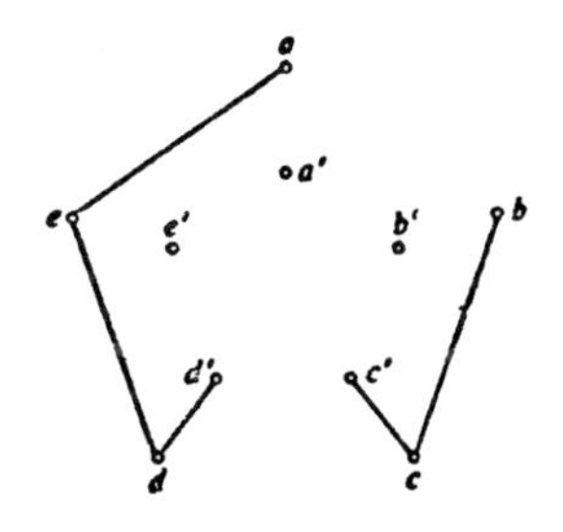

Fig. 19

Second proof. If P were hamiltonian, then we could obtain a 3-coloring of its edges by coloring the edges of a hamiltonian cycle alternatively with two colors, and coloring the remaining edges with the third color. But we saw in the preceding section that there is no 3-coloring of the edges of P, and it follows that P is not hamiltonian. □

Although P is not hamiltonian, it is 'nearly hamiltonian' in a number of senses. For example, it is easy to see that P has a hamiltonian path--an example of such a path is abcdee'c'a'd'b'. Moreover, we can use the symmetry of the Petersen graph to deduce that G has a hamiltonian path starting from any vertex. In fact, since there are automorphisms taking any path of length 3 to any other (see Section 3), every path of length 3 can be extended to a hamiltonian path. However, the corresponding statement for paths of length 4 is false, as may be seen by considering the path aa'd'dc.

It is also easy to see that P is hypohamiltonian, that is, it is not hamiltonian but the removal of any vertex leaves a hamiltonian graph. As before, it is sufficient to verify this for a single vertex. It can be proved that hypohamiltonian graphs or order n exist for all but finitely many values of n, and that the hypohamiltonian graph of smallest order is unique and is the Petersen graph. For more information about hypohamiltonian graphs see, for example, [6].

The Petersen graph has also played a prominent role in connection with two other concepts related to hamiltonian graphs. A graph G is hamiltonian-connected if each pair of vertices of G can be joined by a hamiltonian path, and is 1-hamiltonian if G is hamiltonian and the removal of any vertex (and its incident edges) leaves a hamiltonian graph. At one time it was thought that every graph possesses either both or neither of these properties. To see that this is not the case, consider the two graphs in Figure 20. The first of these is obtained by joining a new vertex to every vertex of P, and the second is obtained by deleting one vertex from P and joining the resulting vertices of degree 2. It is not difficult to verify that the first graph is hamiltonian-connected but not 1-hamiltonian, whereas the second graph is 1-hamiltonian but not hamiltonian-connected.

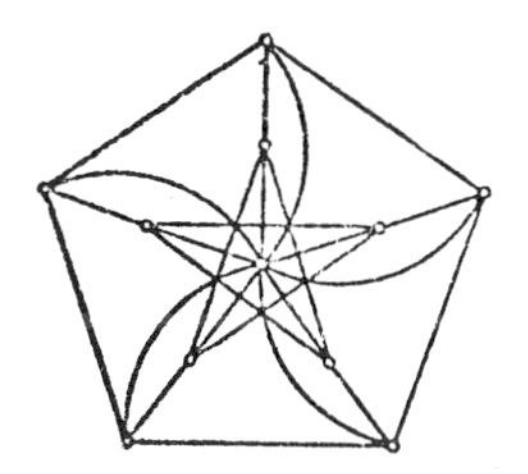
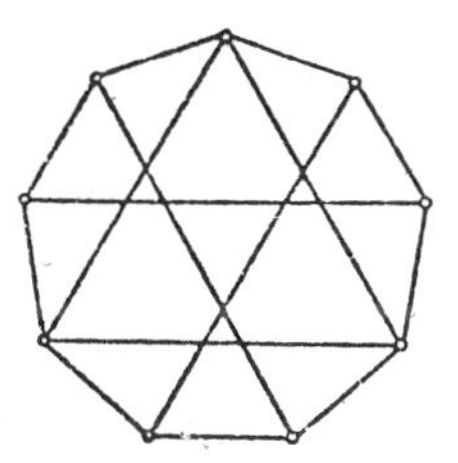

Fig. 20

We conclude this section with one further result which illustrates how very close P is to being hamiltonian. Jackson [21], proved that a 2-connected k-regular graph of order n is hamiltonian if $n \leq 3k$. That 3k cannot be replaced by any larger integer is illustrated by P for which $n = 3k + 1$. In fact, Jackson was able to prove the remarkable result that P is the only 2-connected k-regular non-hamiltonian graph of order $3k + 1$.

6. TOPOLOGICAL PROPERTIES

We begin this section by recalling two famous characterizations of planar graphs, given by Kuratowski [24] and Wagner [38]:

Kuratowski's Theorem. A graph is planar if and only if it contains no subgraph homeomorphic to K_5 or $K_{3,3}$.

Wagner's Theorem. A graph is planar if and only if it contains no subgraph contractible to K_5 or $K_{3,3}$.

Before using these results to prove that P is not planar, we make some further historical remarks. We have already seen that P disproves Tait's statement that every 3-regular bridgeless graph is 1-factorable. Tait also asserted that the graph of every 3-regular polyhedron is 1-factorable. Since a polyhedron is essentially a 3-connected planar graph, as proved by Steinitz [35] in 1922, Tait's assertion concerns the 1-factorability of 3-regular 3-connected planar graphs. As É. Goursat [18] pointed out in 1898, P is not a counter-example to Tait's assertion, since it fails to be planar. But there is a close connection between the 1-factorability of 3-regular planar graphs and the four-color problem on the coloring of maps, and Tait's assertion was eventually proved in 1976 when the four-color problem was settled by Appel and Haken (see, for example, [2]).

We now give three proofs establishing the nonplanarity of the Petersen graph:

Theorem 8. The Petersen graph is not planar.

First proof. If the Petersen graph were embedded in the plane, then the number of regions would be given by Euler's polyhedral formula ($v - e + r = 2$)—namely, $10 - 15 + r = 2$, so that $r = 7$. But P has girth 5, and so each region would be bordered by at least five edges. Since each edge must lie on the boundary of two regions, it follows that $5r \leq 30$, producing the required contradiction. □

Second proof. If we remove the two horizontal edges from P as shown in Figure 21, then the resulting graph is homeomorphic to $K_{3,3}$. The result now follows immediately from Kuratowski's Theorem. □

Third proof. If we contract the five spokes of P, then the resulting graph is K_5. The result now follows from Wagner's Theorem. □

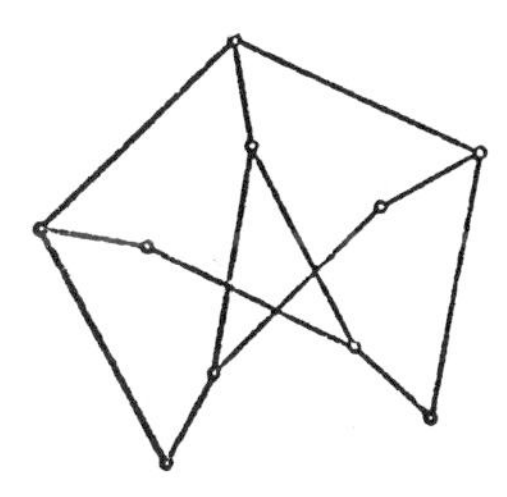

Fig. 21

Since S^P is nonplanar, it cannot be embedded on the surface of a sphere S^o and it is natural to investigate the embeddability of P on other orientable surfaces. In Figure 22 we present an embedding (necessarily a 2-cell embedding) of P on the torus S^1; as usual, we have represented the torus as a rectangle with opposite sides identified. It can be shown that P has 2-cell embeddings on S^1, S^2 and S^3, but on no other orientable surfaces. The Petersen graph can also be embedded on non-orientable surfaces. An example of this is given in Figure 23, which presents an embedding of P on the projective plane.

The nonplanarity of P implies that any drawing of it in the plane must involve a number of crossings, and

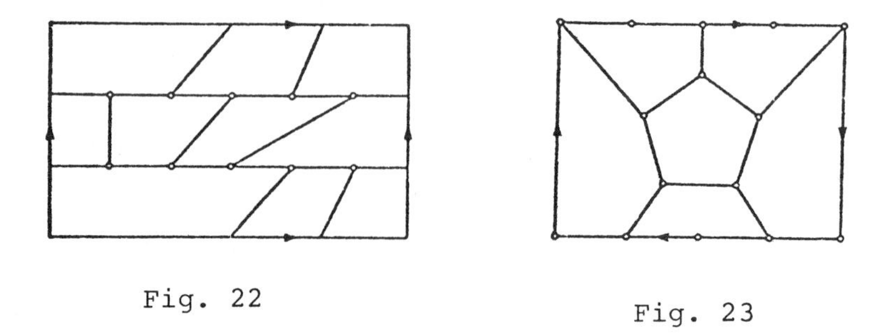

Fig. 22

Fig. 23

it is natural to investigate drawings involving the least such number (the crossing number of the graph). It turns out to be impossible to draw P with only one crossing, but a drawing involving two crossings is possible (see Figure 24). It follows that P has crossing number 2.

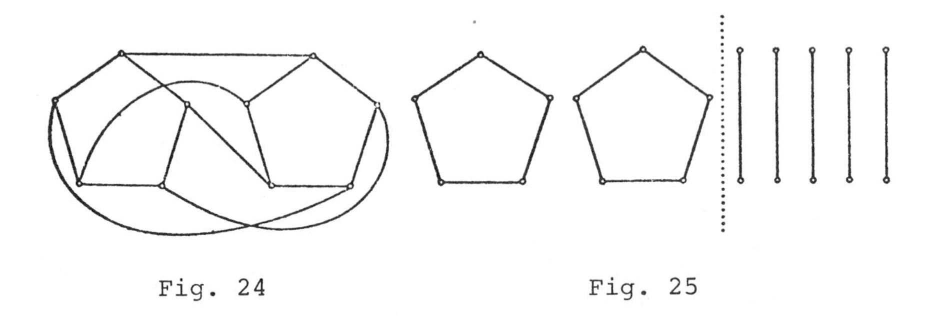

Fig. 24

Fig. 25

Another measure of the nonplanarity of a graph G is its thickness, defined as the least number of planar subgraphs into which G can be split so that a partition of the edge-set of G results. The two subgraphs of P shown in Figure 25 imply that P has thickness 2.

7. GENERALIZATIONS

In this section we describe some of the ways in which the

Petersen graph has been generalized. The resulting graphs are frequently of considerable interest in their own right, and their properties do not always correspond to those of P itself. Further information can be found in [16, Chapter 7] and [8, Chapter 23].

7.1. Cages

We have already defined an r-cage to be a 3-regular graph of smallest order with girth r, and we have seen that the only 3-cage, 4-cage and 5-cage are K_4, $K_{3,3}$, and P, respectively. It can be shown that r-cages exist for all $r \geq 3$, and that there is only one such cage for r = 3,4,5,6,7 and 8. The 6-cage, 7-cage and 8-cage are known, respectively, as the Heawood graph, the McGee graph and the Tutte-Coxeter graph, and are illustrated in Figure 26.

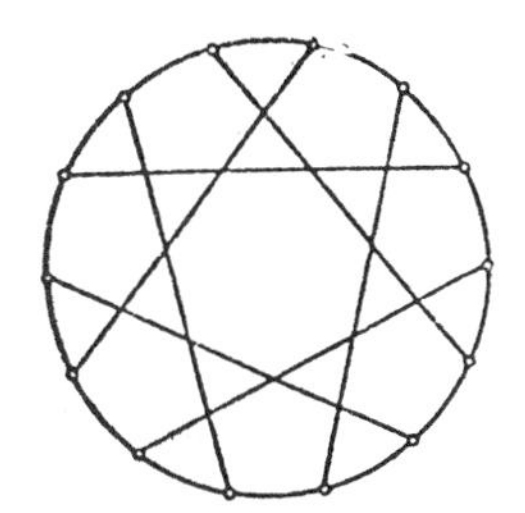
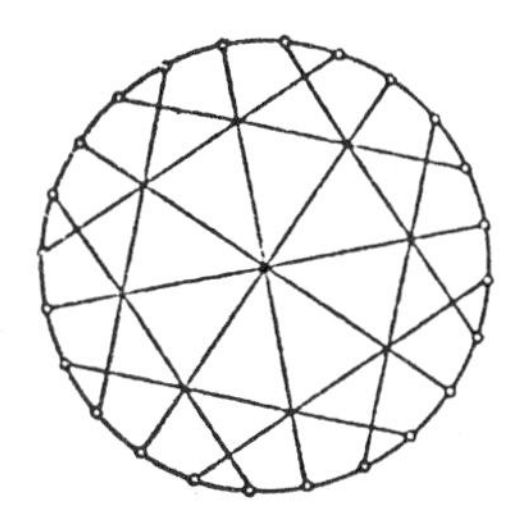
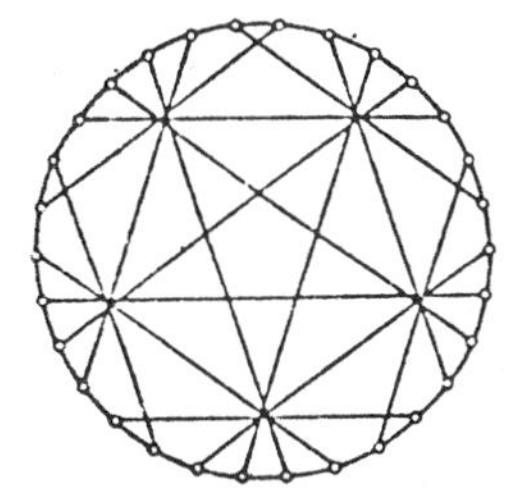

Fig. 26

More generally, we define a (k,r)-cage to be a k-regular graph of smallest order with girth r (so that our previously-defined r-cages are now to be thought of as (3,r)-cages). It is easy to see that the only (2,r)-cage is the r-gon, the only (k,3)-cage is the complete graph K_{k+1}, and the only (k,4)-cage is the complete bipartite graph $K_{k,k}$. It can be shown (see [8, Chapter 23]) that if $k-1$ is a prime power, then the $(k,6)$-cage is the point-line incidence graph of a projective plane of order $k-1$, and that the $(k,8)$-cages and $(k,12)$-cages can be obtained from projective geometries in four and six dimensions. Drawings of some (k,r)-cages with $k > 3$ can be found in [42] and [10, Appendix 3]. We content ourselves with a single example--the $(4,5)$-cage of N. Robertson (Figure 27).

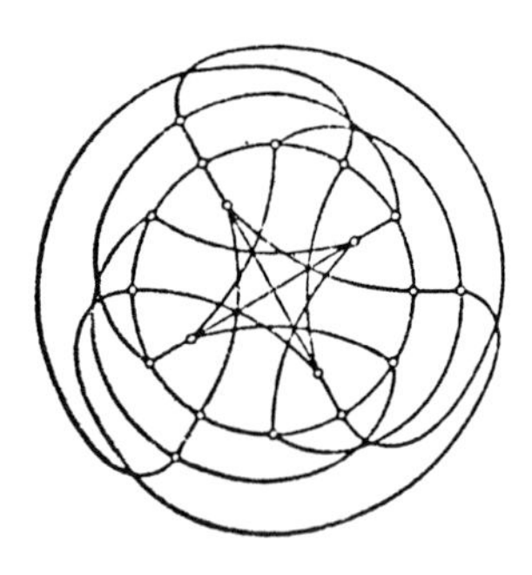

Fig. 27

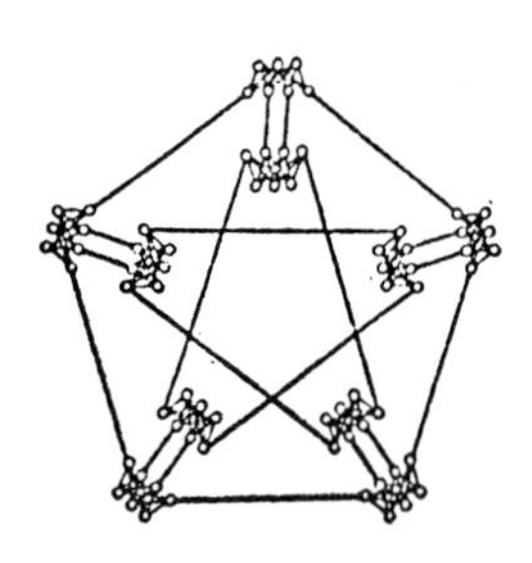

Fig. 28

7.2 Meredith's graphs

We have seen that P is a 3-regular 3-connected non-hamiltonian graph of order 10 whose edges cannot be colored with just three colors. Meredith [26] has generalized this by constructing, for each $k > 3$, a k-regular k-connected non-hamiltonian graph G_k of order $20k-10$. This graph is obtained by replacing each vertex of the Petersen graph by a copy of the complete bipartite graph $K_{k,k-1}$, joined as indicated in Figure 28 which represents the case $k = 4$. It can be shown that if r is the integer nearest to $k/3$ (so that $k = 3r$ or $3r \pm 1$), then the edges of G_k can be colored with k colors only when r is even.

7.3. The Odd Graphs

We have seen (Figure 6) that P may be obtained by taking the ten 2-subsets of {1,2,3,4,5} and joining those pairs of subsets that are disjoint. More generally, the <u>odd graph</u> O_k is obtained by taking as vertices the (k-1)-subsets of $\{1, 2, \dots, 2k-1\}$ and joining those pairs of subsets that are disjoint. It is easy to see that O_2 is simply K_3 and that O_3 is P; a drawing of O_4 appears in Figure 29 (see [27]). Note that the graph O_k is a k-regular graph which has odd order only if k is a power of 2.

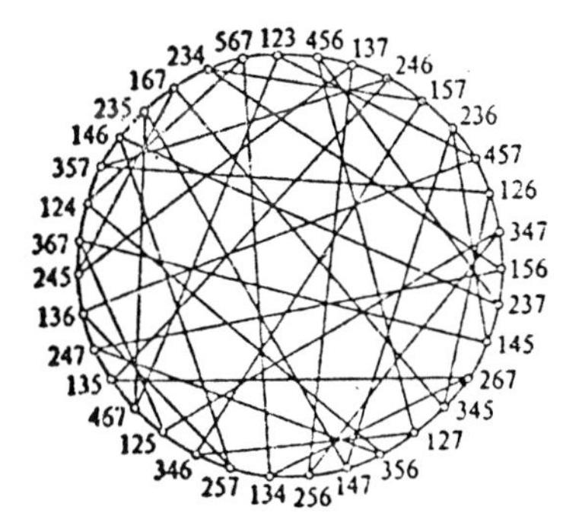

Fig. 29

The odd graphs were first studied in a chemical context under the name of 'k-valent halved combination graphs' (see [4]), but they also appeared in less scientific contexts. For example, Biggs [7] posed the following problem: 'Eleven footballers play five-a-side matches with the eleventh man as referee, and each possible choice of teams and referee play exactly one match. Is it possible to schedule all of the 1386 games in such a way that each individual team plays its six games on six different weekdays?'

By taking the various teams as the vertices of the odd graph 0_6, joining two vertices whenever the corresponding teams play each other, and representing the six weekdays by colors, it is easily seen that a scheduling of the games is possible if and only if the edges of 0_6 can be colored with six colors.

In facet, the edges of 0_6 can be colored with six colors, so the football matches can be scheduled as required. This was established by showing that 0_6 can be split into three disjoint hamiltonian cycles, each of which uses two colors. It is not known in general how many colors are needed to color the edges of 0_k, and which odd graphs are hamiltonian. The present state of knowledge on these questions is summarized in the following table:

k	2	3	4	5	6	7	8
number of colors needed to color the edges of O_k	3	4	5	5	6	7 or 8	9
is O_k hamiltonian?	yes	no	yes	yes	yes	yes	yes

7.4. The Generalized Petersen Graphs

The Petersen graph is obtained by taking an outer 5-cycle, a spoke incident with each vertex of the cycle pointing inward, and an inner 5-cycle joined to the free ends of every second spoke. M.E. Watkins [39] has defined the <u>generalized Petersen graph</u> P(n,k) to consist of an outer n-cycle, n spokes pointing inward, and an inner n-cycle (or a collection of disjoint cycles of total length n) joined to the free ends of every kth spoke. Thus P is the graph P(5,2), and P(9,4) is the graph shown in Fig.30.

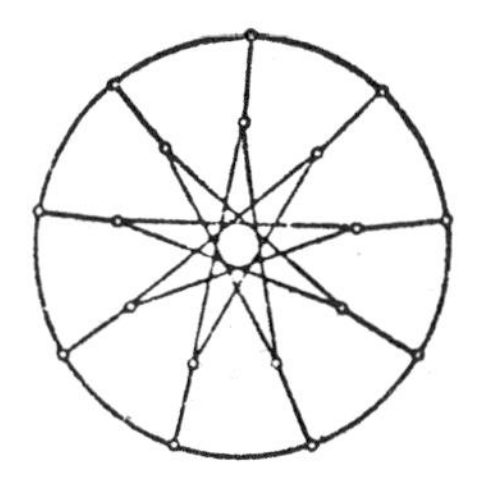

Fig. 30

It can be shown that, with the single exception of P, the edges of $P(n,k)$ can be colored with just three colors.

As with the odd graphs, it is of interest to determine which generalized Petersen graphs are hamiltonian. Alspach [1] has shown that $P(n,d)$ is hamiltonian unless

(i) it is isomorphic to $P(n,2)$, where $n \equiv 5$ (modulo 6),

or (ii) $n = 4r$ and $k = 2r$, where $r \geq 2$.

7.5. Snarks

In recent years, much attention has been paid to the search for bridgeless 3-regular graphs whose edges cannot be colored with three colors. We have already seen in Section 4 that no planar graphs can have this property since such graphs would provide counter-examples to the four-color theorem. The search is therefore directed towards nonplanar graphs of this kind. Because such graphs seem to be difficult to find, Martin Gardner, in a delightful popular article on the subject [17], christened them snarks, after Lewis Carroll's The Hunting of the Snark. In order to avoid trivial cases, we assume that all snarks under consideration have girth 5 or more.

Until recently only four snarks were known. The smallest of these is P, followed by the Blanuša snark

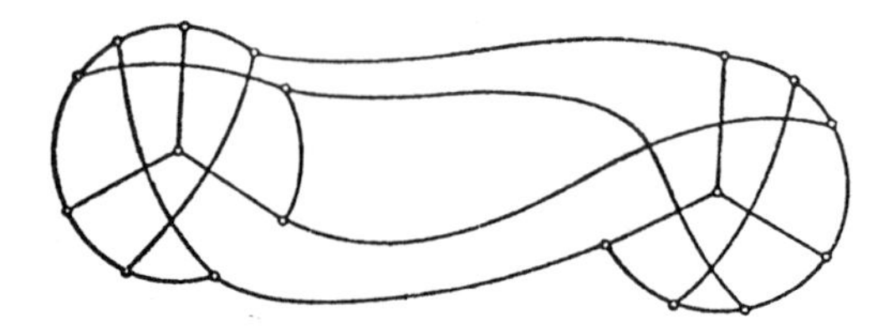

Fig. 31

of order 18 (see Figure 31), the Descartes snark of order 210, and the Szekeres snark of order 50. In 1975 the art of snark-hunting took a dramatic turn when Isaacs [20] described two infinite families of snarks, one of which essentially included the known snarks, and the other of which was completely new. This second family, known as the 'flower snarks', may be obtained by replacing a vertex of the Petersen graph by a triangle and drawing the result as in Figure 32; by replacing the three 'petals' by five, seven, nine,... petals, we obtain the required family. Isaacs also obtained one extra snark—the 'double-star snark'—which does not belong to either family (see Figure 32). (These drawings of the flower snarks and the double-star snark are not those given by Isaacs, but are due to U.A. Celmins and E.R. Swart.) More recently, several new families of snarks have been obtained by a number of people. Details of these snarks are given in [13].

7.6. Tutte's Conjecture

We conclue this section with a conjecture of W.T. Tutte [37] which may help to explain why so much attention has been paid to the Petersen graph and its various generalizations. Every known bridgeless 3-regular graph whose edges cannot be colored with three colors 'contains' the

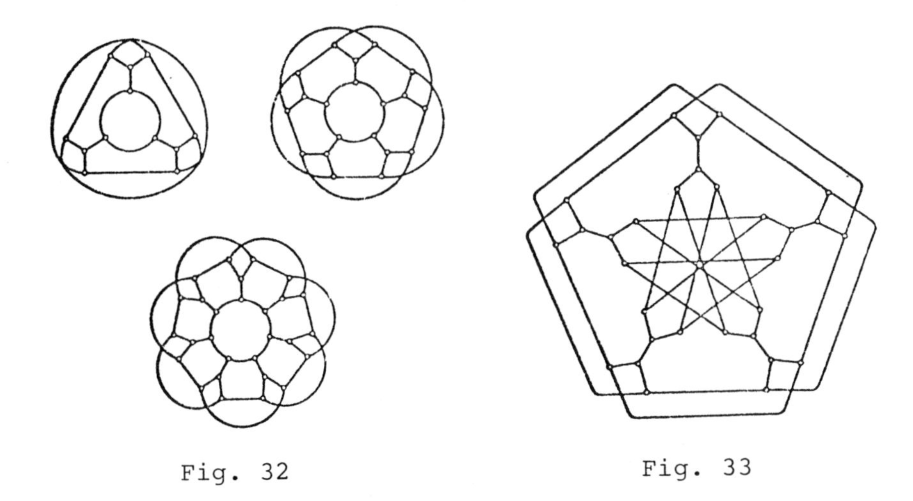

Fig. 32 Fig. 33

Petersen graph. Tutte has conjectured that this is always the case, and his conjecture may be stated in either of the following forms:

Tutte's Conjecture. (i) Every bridgeless 3-regular graph whose edges cannot be colored with three colors contains a subgraph homeomorphic to P,
(ii) every bridgeless 3-regular graph whose edges cannot be colored with three colors contains a subgraph contractible to P.

8. TABLE OF REFERENCE

We conclude with a table of reference summarizing the major parameters associated with the Petersen graph. Some of the terms included here have not been defined earlier in this article, but their definitions can be found in [5] or [19].

parameter	value	parameter	value
number of vertices	10	edge-covering number	5
number of edges	15	arboricity	2
degree of each vertex	3	toughness	4/3
diameter	2	chromatic number	3
radius	2	chromatic polynomial	$\rho(\lambda)$*
girth	5	chromatic index	4
number of 5-cycles	12	genus	1
circumference	9	maximum genus	3
cycle rank	6	crossing number	2
cocycle rank	9	thickness	2
eigenvalues	3;1(×5) -2 (×4)	coarseness	1
number of spanning trees	2000	non-orientable genus	1
connectivity	3	non-orientable maximum genus	6
edge-connectivity	3	automorphism group	S_5
independence number	4	number of automorphisms	120
edge-independence number	5	edge-automorphism group	S_5
covering number	6	strongly-regular parameters	(10,3,0,1)

$$^*\rho(\lambda) = \lambda(\lambda-1)(\lambda-2)(\lambda^7 - 12\lambda^6 + 67\lambda^5 - 230\lambda^4 + 529\lambda^3 - 814\lambda^2 + 775\lambda - 352)$$
$$= \lambda^{10} - 15\lambda^9 + 105\lambda^8 - 455\lambda^7 + 1353\lambda^6 - 2861\lambda^5 + 4275\lambda^4 - 4305\lambda^3 + 2606\lambda^2 - 704\lambda$$

Acknowledgements

The authors gratefully acknowledge the suggestions of Norman L. Biggs, Geoffrey Exoo, S.F. Kapoor, Linda Lesniak-Foster, David Singmaster, Carsten Thomassen and Arthur T. White.

REFERENCES

1. B. Alspach, The classification of hamiltonian generalized Petersen graphs, to appear.
2. K. Appel and W. Haken, Every planar map is four colorable, Bull. Amer. Math. Soc. 82 (1976), 711-712; MR 54#12561.
3. A.T. Balaban, Chemical graphs XIII: Combinatorial patterns. Rev. Roumaine Math. Pures Appl. 17 (1972), 3-16.
4. A.T. Balaban, D. Farcasiu and R. Bănică, Graphs of multiple 1,2-shifts carbonium ions and related systems. Rev. Roumaine Chim. 11 (1966), 1025-1027.
5. M. Behzad, G. Chartrand and L. Lesniak-Foster, Graphs & Digraphs. Prindle, Weber & Schmidt, Boston, (1979); reprinted Wadsworth, Belmont (1981); MR 80f: 05019.
6. J. -C. Bermond, Hamiltonian graphs. Selected Topics in Graph Theory. Academic Press, London (1978), 127-167; MR 81e: 05059.
7. N.L. Biggs, An edge-colouring problem. Amer. Math. Monthly 79 (1972), 1018-1020.
8. N.L. Biggs, Algebraic Graph Theory. Cambridge University Press, London (1974); MR 50#151.
9. N.L. Biggs and D.H. Smith, On trivalent graphs. Bull. London Math Soc. 3 (1971), 155-158; MR 44#3902.
10. J.A. Bondy and U.S.R. Murty, Graph Theory with Applications. American Elsevier, New York (1976); MR 54 #117.
11. R.L. Brooks, On colouring the nodes of a network. Proc. Cambridge Philos. Soc. 37 (1941), 194-197; MR 6-281.

12. P.J. Cameron, Strongly regular graphs. Selected Topics in Graph Theory. Academic Press, London (1978), 337-360; MR 81e: 05059.
13. A.G. Chetwynd and R.J. Wilson, Snarks and supersnarks. Theory and Applications of Graphs. Wiley, New York (1981), 215-241.
14. C. Eisele (ed.), C.S. Peirce: New Elements of Mathematics, Vol. IIIA, Mouton, Den Haag (1976).
15. P. Erdös, F. Harary and W.T. Tutte, On the dimension of a graph, Mathematika 12 (1965), 118-122; MR 32 #5537.
16. S. Fiorini and R.J. Wilson, Edge-colourings of Graphs. Pitman, London (1977): MR 58#27599.
17. M. Gardner, Mathematical games. Scientific American 234, (April 1976), 126-130.
18. É. Goursat, Sur le théorème de Tait. Interméd. Math. 5 (1898), 251.
19. F. Harary, Graph Theory. Addison-Wesley, Reading, MA (1969); MR 41#1566.
20. R. Isaacs, Infinite families of non-trivial trivalent graphs which are non Tait colorable. Amer. Math. Monthly 82 (1975), 221-239, MR 52#2940.
21. B. Jackson, Hamilton cycles in regular 2-connected graphs. J. Combin. Theory (B), to appear.
22. A. B. Kempe, A memoir on the theory of mathematical form. Phil. Trans. Roy. Soc. London 177 (1886), 1-70.
23. A. Kowalewski, Topologische Deutung von Buntordnungsproblemen. Sitzungsber. Akad. Wiss. Wien (Abt IIa) 126 (1917), 963-1007.
24. K. Kuratowski, Sur le problème des courbes gauches en topologie. Fund. Math. 15 (1930), 271-283.
25. K. Menger, Zur allgemeinen Kurventheorie. Fund. Math. 10 (1927), 96-115.

26. G.H.J. Meredith, Regular n-valent, n-connected, non-Hamiltonian, non-n-edge-colourable graphs. J. Combin. Theory (B) 14 (1973), 55-60; MR 47#65.

27. G.H.J. Meredith and E.K. Lloyd, The footballers of Croam, J. Combin. Theory (B) 15 (1973), 161-166; MR 48#49.

28. U.S.R. Murty, On critical graphs of diameter 2. Math. Mag. 41 (1968), 138-140; MR 38#74.

29. J. Petersen, Methods and Theories for the Solution of Problems of Geometrical Construction (Danish). Andr. Fred. Host. and Son, Copenhagen, (1866, 1879). Reprinted in English in String Figures and Other Monographs, Chelsea, New York (1950).

30. J. Petersen, Die Theorie der regulären Graphs. Acta Math. 15 (1891), 193-200.

31. J. Petersen, [Sur le théorème de Tait], Interméd. Math. 5 (1898), 225-227.

32. J. Petersen, [Réponse à question 360], Interméd. Math. 6 (1899), 36-38.

33. A.J. Schwenk, [An elementary problem], to appear.

34. A.J. Schwenk and R.J. Wilson, On the eigenvalues of a graph. Selected Topics in Graph Theory. Academic Press, London (1978), 307-336; MR 81e: 05059.

35. E. Steinitz, Polyeder und Raumentilungen, Encyklopädie der Math. Wiss. IIIAB12 (1922), 1-139.

36. P.G. Tait, Listing's Topologie. Phil. Mag. 17 (1884), 30-46 = Sci. Papers, Vol. 2, Cambridge Univ. Press, Cambridge (1900), 85-98.

37. W.T. Tutte, On the algebraic theory of graph colorings. J. Combin. Theory 1 (1960), 15-50; MR 33#2573.

38. K. Wagner, Über eine Eigenschaft der ebenen Komplexe. Math. Ann. 144 (1937), 570-590.

39. M.E. Watkins, A theorem on Tait colorings with an application to the generalized Petersen graphs. J. Combin. Theory 6 (1969), 152-164; MR 38#4360.

40. H. Whitney, Congruent graphs and the connectivity of graphs. Amer. J. Math. 54 (1932), 150-168.

41. R.J. Wilson, Introduction to Graph Theory, 2d ed. Longman, Harlow, Essex (1979); MR 80g: 05028.

42. P.-K. Wong, Cages--a survey. J. Graph Theory 6 (1982), 1-22.

SEMIREGULAR FACTORIZATIONS OF REGULAR GRAPHS

Hiroshi Era

Department of Mathematical Science
Tokai University
Hiratsuka, Japan

ABSTRACT. A k-semiregular factor, or a (k,k+1)-factor, of G is a factor in which the degree of each point is k or k+1. Then G is k-semiregular factorable if it has a factorization into k-semiregular factors. Our purpose is to give a proof of the following theorem, which proves a conjecture of Akiyama. For each positive integer k there exists an integer $\Phi(k)$ such that for all $r \geq \Phi(k)$, every r-regular graph is k-semiregular factorable.

1. INTRODUCTION

All definitions and notation not presented here can be found in [2]. Let V(G) and E(G) denote the point set and the line set of a graph G, respectively.

Let k be a positive integer. In a k-semiregular graph (or a (k,k+1)-graph) G, each point v has degree k or k+1. A k-semiregular graph G is called k^+-semiregular if G has only one point of degree k+1, and is called $(k+1)^-$-semiregular if G has only one point of degree k. In these cases the point v of degree k+1 (or k) is called the s-point of the k^+-semiregular or $(k+1)^-$-semiregular graph G.

A *factor* of G is a spanning subgraph of G. A k-regular factor is called a *k-factor*. If a factor of G is a k-semiregular subgraph it is called a *k-semiregular factor* (or a *(k,k+1)-factor*) of G. A k-semiregular factor is *minimal* if for any line e of H, H-e is no longer a k-semiregular factor. The union of k-factors (or k-semiregular factors) is called a *k-factorization* *(k-semiregular factorization)* of G and G itself is *k-factorable* (*k-semiregular factorable*).

We will not distinguish a line subset $S \subset E(G)$ from the subgraph $\langle S \rangle$ induced by S. For example, we represent $e \in H$ if e is a line of a subgraph H. An *n-coloring* of the line set of G is a decomposition of the line set $E = E(G)$ into n k-semiregular factors,

$$E = E_1 \cup E_2 \cup \cdots\cdots \cup E_n ,$$

where each E_i is a k-semiregular factor of G. A line e is said to have color i when $e \in E_i$. A point v is *saturated* by the color i if the degree of v in E_i is k+1.

2. LEMMAS

We use the following lemmas.

Lemma A. (Petersen [3]) A graph G is 2-factorable if and only if G is 2n-regular for an integer $n \geq 1$.

Lemma B. (Akiyama, Avis, Era [4], Akiyama [1]) For any integer $r \geq 1$, every r-regular graph is 1-semiregular factorable.

Lemma C. Let G be a graph with a 1-semiregular factor H. If H is minimal, then every component of H is P_2 or P_3.

The proof of this lemma is so immediate so that we omit it.

Lemma D. (Behzad, Chartrand [5]) Let G be a graph with maximum degree $\Delta(G) = r$. Then there exists an r-regular graph which contains G as an induced subgraph.

We use these four Lemmas A, B, C and D freely and without any mention in the proof of the theorem below.

A family of nonempty sets $S_1, S_2, \ldots, S_n$ has <u>a system of distinct representatives</u> if there exists a set of distinct elements $\{s_1, s_2, \ldots, s_n\}$ where $s_i \in S_i$, $i = 1, 2, \ldots, n$.

Lemma E. (Hall [6]) A family of nonempty sets $S_1, S_2, \ldots, S_n$ has a system of distinct representatives if and only if any union of any k sets S_i has at least k elements for each k, $1 \le k \le n$.

3. THE PROOF OF THE THEOREM

We prove the following theorem which establishes the conjecture of Akiyama [1].

Theorem 1. For any positive integer k there exists an integer $\Phi(k)$ such that for all $r > \Phi(k)$, every r-regular graph is k-semiregular factorable. Moreover, the smallest $\Phi(k)$ is not greater than the value of a quadratic polynomial in k.

<u>Proof</u>. We divide the proof into two cases depending on the pariety of k, each of which is claimed as a

proposition. We fix k as an arbitrary positive integer, and m as a non-negative integer such that $0 \le m \le k-1$, throughout the proof.

Proposition 1. Any r-regular graph G is $2k$-semiregular factorable if r is an integer which satisfies one of the following conditions.

(1) $r = 2kn + 2m$, where n is a positive integer and $n \ge 3m$.

(2) $r = 2kn + 2m + 1$, where n is a positive integer and $n \ge 4m + 1$.

Proof. First, assume that r satisfies (1).
Let $F_1, F_2, \ldots, F_m$ be any m mutually disjoint 2-factors of G and let $H = G - \bigcup_{t=1}^{m} F_t$. Since H is $2kn$-regular, H can be decomposed into n $2k$-factors $H_1, H_2, \ldots, H_n$. The operation for coloring the lines consists of two parts. To begin, we color all lines of H_i by the i-th color, $i = 1, 2, \ldots, n$.

Since any connected component of F_t, $t=1,2,\ldots,m$, is a cycle, all lines of F_t can be colored by the three colors numbered $3t-2$, $3t-1$ and $3t$ in such a way that any two adjacent lines in F_t have different colors each other.

Then we color all the lines of F_t according to the condition described above, $t = 1, 2, \ldots, m$. Note that in so coloring the F_t we do not use any colors other than the n colors used in coloring the H_i, since $n \ge 3m$ by (1). It is easily seen that these colorings combine to give an n-coloring of G, completing the proof of case (1).

Second assume that r is an integer satisfying (2). Let F be a 1-semiregular factor of G and let $H = G - F$.

Then H is an $(r-1)$-semiregular graph and $\Delta(H) = r-1$. Embed H into an $(r-1)$-regular graph $\tilde{H}$ so that H is an induced subgraph of $\tilde{H}$. Noting that $r-1 = 2kn + 2m$, we can find mutually disjoint m 2-factors $\tilde{F}_1, \tilde{F}_2, \ldots, \tilde{F}_m$ of $\tilde{H}$. Put $\tilde{K} = \tilde{H} - \bigcup_{t=1}^{m} \tilde{F}_t$, then K is $2kn$-regular and can be decomposed into n $2k$-factors $\tilde{K}_1, \tilde{K}_2, \ldots, \tilde{K}_n$. Put $K_i = \tilde{K}_i \cap H$, $i = 1,2,\ldots,n$.

In the following we give an n-coloring of the line set of G. The operation consists of four parts:

(i) We color all lines of K_i by the i-th color, $i = 1,2,\ldots,n$.

From the definitions of H and $\tilde{H}$ it is immediate that for any point v of G,

$$\deg_H v \geq \deg_{\tilde{H}} v - 1.$$

Hence,

$$2k-1 \leq \deg_{K_i} v \leq 2k, \quad i = 1,2,\ldots,n.$$

In particular, if

$$\deg_{K_i} v = 2k-1 \tag{1-1}$$

for some i, then

$$\deg_{K_j} v = 2k \ (j \neq i) \tag{1-2}$$

and

$$\deg_F v = 2. \tag{1-3}$$

(ii) For each point v that satisfies (1-1) for some i, we color the two lines of F to which v is incident by the i-th color.

By (ii), some component of F which is isomorphic to P_3 is colored. For the other components of F isomorphic to P_2 and P_3 which remain uncolored, we now perform:

(iii) We color one line of each component by an arbitrary color (say the first color).

After performing (i), (ii) and (iii), for every point v of G the set of all lines incident to v satisfies the next condition (a).

(a) There is just one color that is saturated at v and any other color is incident to v in exactly $2k$ lines.

Now we assign colors to all lines belonging to $\tilde{F}_1,\ldots,\tilde{F}_m$, which remain uncolored. When $m = 0$, we need no additional operations.

In case $m \geq 1$, a procedure to color those lines is stated in several steps as follows.

[I] The lines of $\tilde{F}_1$ are denoted by $e_1, e_2, \ldots, e_\ell$. We color these lines one by one, in order, by a color which is not saturated at two endpoints of each line. When we colored the first j lines ($j = 0,1,2,\ldots$), the sum of the numbers of colors which are saturated at each endpoint of the $(j+1)$-st line is, by condition (a), at most four, since $\tilde{F}_1$ is a 2-factor. By the assumption $n \geq 4m + 1$ in (2), there exists a color which is not saturated in G at the points of e_{j+1}. Hence we can color all lines $e_1, e_2, \ldots, e_\ell$ in such a way that the number of lines of each color at each point of G does not exceed $2k + 1$.

[II] We assume that all the lines of $\tilde{F}_1, \tilde{F}_2, \ldots, \tilde{F}_{t-1}$ are colored adequately ($2 \leq t \leq m$), and the next condition (b) is satisfied at each point v of G.

(b) In G, the number of saturated colors at v is at most $2t - 1$ and any other color is incident to v in at most $2k$ lines.

The lines of $\tilde{F}_t$ are denoted by $e_1, e_2, \ldots, e_\ell$. We color each line one by one in order. When we reach e_{i+1} after coloring e_i to e_j $(j=0,1,2,\ldots)$, the sum of the numbers of saturated colors at each point of e_{j+1} is, by (b), at most $2\{(2t-1)+1\} = 4t$ in G, since $\tilde{F}_t$ is a 2-factor. By the assumption $n \geq 4m+1 > 4t$ in (2), there exists a color which is not saturated in G at both points of e_{j+1}. Thus we can color all the lines $e_1, e_2, \ldots, e_\ell$ in such a way that the number of lines of each color at any point of G does not exceed $2k+1$. Moreover the next condition (c) holds at each point v of G.

(c) In G, the number of saturated colors at v is at most $2(t+1)-1$ and any other color is incident to v in at most 2k lines.

By the procedure described in [I] and [II],

(iv) We can color all the lines of $\tilde{F}_1, \tilde{F}_2, \ldots, \tilde{F}_m$ so that at each point v of G, every color is incident to v in 2k or $2k+1$ lines in G.

Finally, we color the remaining lines of F each of which is in one of the P_3 components. For this operation it is sufficient to note the following fact. Let uv be such a line, $\deg_F u = 2$ and $\deg_F v = 1$. Then there must be an $\tilde{F}_i$ which has a line incident to u and not belonging to G, and also the operation (iii) does not increase the number of colors at v, so that we can conclude that each of u and v has at most 2m saturated colors. Hence,

(v) We can color the remaining lines of F.

By (i) to (v) an n-coloring of G is given, by which the proof of case (2) is accomplished.

To prove the other assertion, that is, every r-regular graph is $(2k-1)$-semiregular factorable for a sufficiently large integer r, we need some more lemmas.

Lemma 1. Let G with q lines be a connected $2k$-regular $(k \geq 1)$ graph, and let uv be any line of G.

(a) If q is even, then G can be decomposed into two k-factors G_1, G_2.

(b) If q is odd, then G can be decomposed into two factors G_1, G_2 such that

  (1) G_1 and G_2 are k^--semiregular and k^+-semiregular, respectively,

  (2) the point u is the common s-point of G_1 and G_2,

  (3) $uv \in G_2$.

Proof. Let $\varepsilon: u_0, u_1, \ldots, u_q, u_0$ be an eulerian trail of G. We may choose u and v for u_0 and u_1, respectively. Then G_1 and G_2 are defined as follows:

$$u_{2i+1}u_{2i} \in G_1, \quad u_{2i}u_{2i+1} \in G_2, \quad i = 0, ., \ldots,$$
$$\text{and} \quad u_q u_0 \in G_2 .$$

It is easy to see that G_1 and G_2 satisfy the condition of the lemma in either case (a) or (b). □

Lemma 2. Let G have $n(k-1)$-semiregular factors $H_1, H_2, \ldots, H_n$ and a minimal 1-semiregular factor F, $n \geq 1$, $k \geq 3$. Let the components of each H_i be $C_{i,1}, C_{i,2}, \ldots, C_{i,\alpha(i)}$ and of F be $D_1, D_2, \ldots, D_\beta$. If $k \geq 3n$, then there exists an injection ψ from the set $C = \{C_{i,j} \mid i = 1,2,\ldots,n;\ j = 1,2,\ldots,\alpha(i)\}$ to the set $D = \{D_j \mid j = 1,2,\ldots,\beta\}$ such that $V(C_{i,j}) \cap V(\psi(C_{i,j})) = \emptyset$.

Proof. Put $\tilde{C}_{i,j} = \{D_h \in D \mid V(D_h) \cap V(C_{i,j}) \neq \emptyset\}$. Consider a union of any $t \geq 2$ of the $\tilde{C}_{i,j}$, say, $\bigcup_{(i,j)} \tilde{C}_{i,j}$. After we show that $|\bigcup_{(i,j)} \tilde{C}_{i,j}| \geq t$, the proof is accomplished at once by using Lemma E.

Since $|V(C_{i,j})| \geq k$, and each point v of G belongs to at most n $C_{i,j}$s, the inequality

$$|\bigcup_{(i,j)} V(C_{i,j})| \geq \frac{tk}{n}$$

holds. By the assumption $k \geq 3n$, the above inequality is equivalent to the following:

$$|\bigcup_{(i,j)} V(C_{i,j})| \geq 3t \tag{2-1}$$

On the other hand, by the assumption that F is minimal, the number of points of D_h is bounded by

$$2 \leq |V(D_h)| \leq 3, \quad h = 1,2,\ldots,\beta \tag{2-2}$$

Using (2 1) and (2-2) it is easily shown that

$$|\bigcup_{(i,j)} \tilde{C}_{i,j}| \geq \frac{3t}{3} = t\ .$$

This completes the proof of Proposition 1. □

Proposition 2. An r-regular graph G is $(2k-1)$-semi-regular factorable if r satisfies one of the following conditions where $n \geq 2k - 2m - 1$ is positive:

(1) $r = 2kn + 2m$,

(2) $r = 2kn + 2m + 1$.

Proof. First assume that r and k satisfy (1). When $m = 0$ the proposition is imediately derived from

Lemma A. Hence assume that $m \geq 1$. Let F be an arbitrary 2-factor of G and let $H = G - F$. Then H is $(r-2)$-regular. Considering the integers $N = n - 2k + 2m-1$, which is nonnegative by the assumption in (1), and $M = k - m + 1$, H can be decomposed into N $2k$-factors $D_1, D_2, \ldots, D_N$ and M $2(2k-1)$-factors $H_1, H_2, \ldots, H_M$, since $r-2$ can be written as $r-2 = 2(2k-1)M + 2kN$. We denote the components of H_i by $K_{i,1}, K_{i,2}, \ldots, K_{i,\alpha(i)}$, $i = 1,2,\ldots,M$. We obtain a minimal 1-semiregular factor of F on deleting some appropriate lines from F and we denote the resulting components by $F_1, F_2, \ldots, F_\beta$. Using Lemma 2 we get an injection ψ from the set $\{K_{i,j} : i = 1,2,\ldots,M;$ $j = 1,2,\ldots,\alpha(i)\}$ to the set $\{F_j : j = 1,2,\ldots,\beta\}$, where the hypothesis of the lemma is assured by the inequality $2(2k-1) + 1 \geq 3M$, which holds under the assumption $1 \leq m \leq k-1$. Put $\psi(k_{i,j}) = F_{i,j}$. From Lemma 1, $K_{i,j}$ can be decomposed into two subgraphs $J(i,j,1)$ and $J(i,j,2)$ such that

(a) If $|E(K_{i,j})|$ is even, then $J(i,j,1)$ and $J(i,j,2)$ are both $(2k-1)$-regular.

(b) If $|E(K_{i,j})|$ is odd, then

1. $J(i,j,1)$ and $J(i,j,2)$ are $(2k-1)^-$-semiregular and $(2k-1)^+$-semiregular, respectively, and
2. the common s-point of $J(i,j,1)$ and $J(i,j,2)$ belongs to $F_{i,j}$.

Now we construct an n-coloring of G, in five steps.

(i) We color all lines of $J(i,j,t)$ by color number $2(i-1) + t$, $i = 1,2,\ldots,M$; $j = 1,2,\ldots,\alpha(i)$, $t = 1,2$.

(ii) For all $J(i,j,1)$ which are $(2k-1)^-$-semiregular, we color the one or two lines of $F_{i,j}$ incident to the s-point v_{ij} by color $2i-1$.

(iii) We color the lines of D_j by color $2M+j$, $j-1,2,\ldots,N$.

Note that we use the 2M colors $1,2,\ldots,2M$ in (i) and (ii), and the N colors $2m+1,\ldots,2M+N = n+1$ in (iii).

After performing (i), (ii), and (iii), all the lines remaining uncolored belong to F. Therefore they induce a 1-semiregular subgraph of G, any nonempty connected component of which is a cycle C_p or a path P_s. It can be seen that for each point v of those C_p, after doing (i), (ii), and (iii), any one of the first 2M colors is incident to v in exactly 2k-1 lines. Since $2M \geq 4$,

(iv) We can color all lines of each C_p by using the first 2M colors in such a way that any two adjacent lines have different colors.

Finally consider a path P_s of a component of the uncolored subgraph. It is easily seen that for each point v of P_s, after doing (i) to (iv), the number of saturated colors among $1,2,\ldots,2M$ at v is at most one if v is one of the endpoints of P_s, and zero if v is an inner point of P_s. Therefore

(v) We can color all lines of P_s in such a way that any two adjacent lines have different colors. In particular, the endpoints of P_s can be colored by colors which are not saturated at them.

By (i) to (v) we get an n-coloring of G, completing case (1).

Now we prove the case defined by condition (2). Let F be a minimal 1-semiregular factor of G, $V_1 = \{v \in V(G) : \deg_F v = 1\}$, $V_2 = \{v \in V(G) : \deg_F v = 2\}$, $H = G - F$. Then H is an (r-1)-semiregular graph, and there is an (r-1)-regular graph $\tilde{H}$ which contains H as an induced subgraph. Note that

$$v \in V_1 \text{ implies } \deg_H v = \deg_{\tilde{H}} v$$

and (3-1)

$$v \in V_2 \text{ implies } \deg_H v = \deg_{\tilde{H}} v - 1$$

Consider the integer $N = n + 2m - 2k + 1$, which is non-negative by the assumption in (2), and $M = k - m$, $\tilde{H}$ can be decomposed into N $2k$-factors $\tilde{D}_1, \tilde{D}_2, \ldots, \tilde{D}_N$ and M $2(2k-1)$-factors $\tilde{H}_1, \tilde{H}_2, \ldots, \tilde{H}_M$, since the equality $r - 1 = 2(2k-1)M + 2kN$ holds. We denote the connected components of each $\tilde{H}_i$ by $\tilde{K}_{i,1}, \tilde{K}_{i,2}, \ldots, \tilde{K}_{i,\alpha(i)}$, $i = 1,2,\ldots,M$. Put $K_{i,j} = \tilde{K}_{i,j} \cap H$. By Lemma 2 there exists an injection ψ from the set $\{K_{i,j} \mid i = 1,\ldots,M; j = 1,2,\ldots,\alpha(i)\}$ to the set of the components of F. Put $\psi(K_{i,j}) = F_{i,j}$. Using Lemma 1, each $\tilde{K}_{i,j}$ can be decomposed into two subgraphs $\tilde{J}(i,j,1)$ and $\tilde{J}(i,j,2)$ as follows.

(a) If $|E(\tilde{K}_{i,j})|$ is even, then $\tilde{J}(i,j,1)$ and $\tilde{J}(i,j,2)$ are both $(2k-1)$-regular.

(b) If $|E(\tilde{K}_{i,j})|$ is odd, then

1. $\tilde{J}(i,j,1)$ and $\tilde{J}(i,j,2)$ are $(2k-1)^-$-semiregular and $(2k-1)^+$-semiregular, respectively, and
2. the common s-point v_{ij} of $\tilde{J}(i,j,1)$ and $\tilde{J}(i,j,2)$ belongs to $F_{i,j}$, and
3. a line e of $\tilde{H} = H$ which is incident to v_{ij}, if any, belongs to $\tilde{J}(i,j,2)$.

Put $J(i,j,t) = \tilde{J}(i,j,t) \cap H$.

We now show from the definition of $J(i,j,t)$ and equation (3-1) that each point v of G has such a degree in $J(i,j,t)$.

Case 1. $v \in V_1$ and v is not an s-point of any $\tilde{J}(i,j,t)$.

$$\deg_{J(i,j,t)} v = 2k - 1 \quad \text{for all } (i,j,t) \tag{3-2}$$

Case 2. $v \in V_1$ and v is the s-point of one of the $\tilde{J}(i,j,t)$, say $\tilde{J}(i',j',t')$.

$$\deg_{J(i',j',1)} v = 2k-2, \ \deg_{J(i',j',2)} v = 2k \text{ and}$$

$$\deg_{J(i,j,t)} v = 2k - 1 \text{ for any } (i,j,t) \neq (i',j',t') \tag{3-3}$$

Case 3. $v \in V_2$ and v is not an s-point of any $\tilde{J}(i,j,t)$.

There is at most one triple among the (i,j,t), say (i',j',t'), such that

$$\deg_{J(i',j',t')} v = 2k - 2,$$

and

$$\deg_{J(i,j,t)} v = 2k - 1$$

for any $(i,j,t) \neq (i',j',t')$. (3-4)

Case 4. $v \in V_2$ and v is the s-point of one of the $\tilde{J}(i,j,t)$, say $\tilde{J}(i',j',t')$.

There is at most one triple among the (i,j,t), say (i'',j'',t''), such that

$$\deg_{J(i'',j'',t'')} v = 2k - 2,$$

$$\deg_{J(i',j',1)} v = 2k - 2, \ \deg_{J(i',j',2)} v = 2k-1 \text{ or } 2k$$

and

$$\deg_{J(i,j,t)} v = 2k - 1$$

for any $(i,j,t) \neq (i',j',t'), (i'',j'',t'')$. (3-5)

Note that (i'',j'',t'') is or is not $(i',j',1)$, according to whether $\deg_{J(i',j',2)} v = 2k - 1$ or $2k$.

Now we give an n-coloring to G.

(vi) We color all lines of $J(i,j,t)$ by color $2(i-1)+t$, $i = 1,2,\ldots,M$; $j = 1,2,\ldots,\alpha(i)$, $t = 1,2$. □

In (vi), 2M colors are used.

(vii) We color all lines of D_i by color $2M+i$, $i = 1,2,\ldots,N$. □

The number of colors used in (vi) and (vii) is $2M+N = n+1$. After (vi) and (vii) the uncolored lines are only those of the 1-semiregular factor F. Finally we color those lines of F by taking its components one by one. There are several cases as follows.

Case 1F. Consider a component uv isomorphic to P_2. There are two possibilities. If neither of u and v are s-points of any $\tilde{J}(i,j,t)$, then by (3-2),

(viii) We can color the line uv by an arbitrary color so that the number of lines in each color incident to u and v are 2k or 2k - 1.

On the other hand if v or u is an s-point of some $\tilde{J}(i,j,t)$, then by (3-3),

(ix) Coloring the line uv by color $2i'-1$, both u and v have 2k or 2k-1 incident lines of every color, where i' means the same index as in (3-3).

Case 2F. Consider a component uvw isomorphic to P_3. Now there are three possibilities. First, consider any one of u, v and w to not be an s-point. From (3-2) and (3-4),

(x) Coloring the lines uv and vw by color $2(i'-1)+t$, each one of these three points has 2k or 2k-1 incident lines of every color, where i' means the same index as in (3-4).

Second, consider v as an s-point of some $\tilde{J}(i,j,t)$. From (3-2) and (3-5),

(xi) Coloring the lines uv by color $2(i''-1)+t$ and vw by color $2i'-1$, each of these three points has 2k or 2k-1 incident lines of each color, where i' and i" are the same indexes as in (3-5).

Finally, let u or v be an s-point of some $J(i,j,t)$. From (3-2), (3-3) and (3-4),

(xii) Coloring uv by color $2i'-1$, vw by color $2(i''-1)+t$, each of these three points has degree 2k or 2k-1 in every color, where i' and i" means the same indexes as i' in (3-3) and (3-4) respectively.

It can be seen that (vi) to (xii) give an n-coloring of G, and the proof of Proposition 2 is finished. □

Combining Propositions 1 and 2 with Lemma B results in Theorem 1. □

REFERENCES

1. J. Akiyama, Factorization and Linear Arboricity of Graphs, Doctoral dissertation, Science University of Tokyo (1982).
2. F. Harary, Graph Theory, Addison Wesley, Reading (1969).
3. J. Petersen, Die Theorie der regulären graphs, Acta Math. 15(1891) 193-220.
4. J. Akiyama, D. Avis, H. Era, On a {1,2}-factor of a graph, TRU Math. 16(1980) 97-102.

5. M. Behzad, G. Chartrand and E.A. Nordhaus, Triangles in line-graphs and total graphs, Indian J. Math. 10 (1968) 109-120.

6. P. Hall, On representatives of subsets, J. London Math. Soc. 10(1935) 26-30.

GENERALIZATIONS OF THE TREE-COMPLETE GRAPH RAMSEY NUMBER

R.J. Faudree
C.C. Rousseau
R.H. Schelp

Department of Mathematical Sciences
Memphis State University
Memphis, Tennessee 38152

ABSTRACT. There have been several different generalizations of Chvátal's tree-complete graph ramsey number. A theorem is presented which includes many of these as corollaries, and a comparison is made of this result with some of the known generalizations.

1. INTRODUCTION

Let F and G be (simple) graphs. The ramsey number $r(F,G)$ is the smallest positive integer p such that if each line of the complete graph K_p is colored one of the two colors red or blue, then either the red subgraph contains a copy of F or the blue subgraph contains a copy of G. Surveys including many of the known results of this subject are given in [2] and [11]. Here we restrict ourselves to generalizations of the following simply proved result of Chvátal [8]. Notation throughout follows that given in [12].

Theorem A. If T_n is a tree on n points, then

$$r(K_k,T_n) = (k-1)(n-1) + 1.$$

For several nice generalizations of this theorem, see [1], [3], [5], [9], and [10]. Our objective is to present the following theorem and see that it includes several of these generalizations as special cases.

Theorem 1. Let G be a connected graph on n points and F a fixed graph on p points with chromatic number $\chi = \chi(F)$. Let $s = s(F)$ be the smallest number of points in a color class under any χ-coloring of F. There exist positive constants ε_1 and ε_2 such that when n is sufficiently large, $q(G) \leq n + \varepsilon_1 n^{1/(2p-1)}$ and $\Delta(G) \leq \varepsilon_2 n^{1/(2p-1)}$ imply that

$$r(F,G) = (\chi - 1)(n-1) + s.$$

2. KNOWN GENERALIZATIONS

Before discussing Theorem 1 we present some of the better known generalizations of Theorem A. Many of these results involve graphs which are said to be k-good. A k-good graph is a connected graph G on n points such that $r(K_k,G) = (k-1)(n-1) + 1$. Burr and Erdös in [3] and [4] list most of the graphs known to be k-good. Many of these graphs have "few" lines or have a special "regular" structure. We list some of their results in the next theorem.

Theorem B. Let k be a fixed positive integer.

(i) Let G be a graph obtained from a connected graph by successively subdividing its lines resulting in a graph with a long suspended path (a path in which each point is of degree 2 in G). For p(G) sufficiently large, G is k-good.

(ii) Let G be a graph obtained from a connected graph by successive additions of end lines. For p(G) sufficiently large G is k-good.

(iii) Let t be fixed. For n sufficiently large, P_n^t (the t-th power of path P_n) is k-good.

(iv) Let t be fixed and let G be a graph homeomorphic from a fixed connected graph. For p(G) sufficiently large the graph G^t is k-good.

(v) The wheel W_n is 3-good for $n \geq 6$.

(vi) The subdivision graph $S(K_n)$ is 3-good for $n \geq 8$.

(vii) Let K"(n) denote the graph obtained from K_n by replacing each of its lines by a triangle. For $n \geq 8$, the graph K"(n) is 3-good.

The parameter k appearing in the ramsey number $r(K_k,T_n) = (k-1)(n-1) + 1$ may be interpreted as the chromatic number of K_k. Using this interpretation, a natural definition for a graph to be F-good can be given. For a graph F, letting s(F) be the parameter described in Theorem 1, define a connected graph G as <u>F-good</u> if $r(F,G) = (\chi(F)-1)(p(G)-1) + s(F)$. Note that $(\chi(F)-1)(p(G)-1) + s(F)$ is clearly a lower bound for r(F,G); simply take $(\chi(F)-1)K_{p(G)-1} \cup K_{s(F)-1}$ as the blue subgraph and its complement as the red subgraph in a 2-coloring of the lines of a complete graph on $(\chi(F)-1)(p(G)-1) + s(F)-1$ points.

Several graphs or families of graphs are known to be F-good. We list some representative ones in the following theorem.

Theorem C.

(i) [7]. Let G be an arbitrary nonbipartite graph. For $n > p(G)$, let G_n be the graph consisting of G together with a path of length $n-p(G)$ emanating from one of the points of G. For n sufficiently large, G_n is G_n-good, i.e., $r(G_n, G_n) = (\chi(G_n)-1)(n-1) + s(G_n)$.

(ii) [10]. If $n \geq 3$, $m \geq 6$, and T_n is any non-star tree of order n, then T_n is $(K_m - tK_2)$-good for each t, $0 \leq t \leq [\frac{m}{2}] - 1$.

(iii) [9]. Let G be a connected graph with n points and $q(G) \leq n(1+\varepsilon)$ lines, and let C_{2k+1} be an odd cycle. For ε sufficiently small (depending on k) and n sufficiently large (depending on ε), the graph G is C_{2k+1}-good.

Observe that in several of the results given in Theorem B the ramsey number remained unchanged when the tree T_n of Theorem A was replaced by a connected graph with a "few" additional lines. In Theorem C (i), and (iii), the graphs K_k and T_n of Theorem A were both altered while in Theorem C (ii) only K_k was changed. Effectively a "few" lines are added to the tree T_n while some are deleted from K_k giving a "good" ramsey number. This technique is a nice one in that it provides a means for finding the ramsey number for certain pairs of graphs in terms of well known graphical parameters. Theorem 1 has Theorem C (ii) and (iii) and Theorem B (i) and (ii) as corollaries. In particular well known ramsey numbers for special pairs of graphs are corollaries to Theorem 1. For example, for n large, each of $r(K_m, C_n)$, $r(C_k, C_n)$, $r(C_k, P_n)$, $r(T_k, P_n)$, $r(T_k, C_n)$, $r(K(s_1, s_2, \ldots, s_t), P_n)$, $r(K(s_1, s_2, \ldots, s_t), C_n)$ are given by Theorem 1.

3. THE MAIN RESULT

The details of the proof of the main result, Theorem 1, will not be given here. Rather only the necessary ingredients with a sketch of the proof will be presented. A complete detailed proof will appear elsewhere [6].

The proof depends on a Lemma and Proposition. The Lemma appears in [9].

Lemma 1. Let G be a graph with no isolated points and no suspended path with more than s points such that $q(G) > p(G)$. Then G has at least $\left\{\frac{p(G)}{2s} - \frac{3(q(G)-p(G))}{2}\right\}$ points of degree 1.

Proposition 1. Let G be a graph without isolated points such that $q(G) \leq p(G) + k$. Let $s_1 \leq s_2 \leq \ldots \leq s_t$, $s = \Sigma s_i$, and $\Delta = \Delta(G)$. Then

$$r(K(s_1, s_2, \ldots, s_t), G) \leq (t-1)(p(G)-1) + (3s^2k + 2s^4\Delta)^{s-1}.$$

We present only the ideas of this proof which are the same as those used in proving Theorem 1.

<u>Sketch of Proof</u>. The proof is by double induction on $n = p(G)$ and t. The result is trivial for $t = 1$ and can be easily checked for $n \leq 3s^2k + 2s^4\Delta$ using the estimate $r(K_s, K_n) \leq \min\{n^{s-1}, s^{n-1}\}$. Thus we assume $t > 1$, $n > 3s^2k + 2k^4\Delta$ and that the result holds for smaller values of n or t.

The proof proceeds by considering two cases, the first when G has a "long" suspended path and the second when G has "many" end lines. One of these two cases must occur since otherwise n is found to be smaller than $3s^2k + 2s^4$.

Instead of doing all the arithmetic precisely, we outline the strategy involved in each of the two cases. Throughout the remainder of the proof we assume that the 2-colored K_ℓ, $\ell = (t-1)(n-1) + (3s^2k + 2s^4\Delta)^{s-1}$, contains no red $K(s_1,s_2,\ldots,s_t)$ and no blue G as subgraphs.

If G contains a long suspended path with m vertices, then by induction on n the graph G' obtained from G by shortening the suspended path by one point is a blue subgraph of K_ℓ. Also by induction on t there is a red $H = K(s_1,s_2,\ldots,s_{t-1})$ in K_ℓ which is point disjoint from the blue copy of G. Let P be the suspended path of G' with $m-1$ points.

If a point of H is adjacent in blue to two consecutive points of the path P, then there is a blue G in K_ℓ, so this cannot happen. If a point h of H is adjacent in blue to s points of P, then the successors of the first $s-1$ of these points on P are pairwise adjacent in red. But then these $s-1$ successors together with h form a red K_s in K_ℓ . Hence we can assume that each point of H is adjacent in blue to at most $s-1$ points of the path P. For m sufficiently large ($m \geq s^2$ works), we find there are at least s points of the path P which are adjacent in red to all points of H. This gives a red $K(s_1,s_2,\ldots,s_t)$ in K_ℓ , a contradiction.

If G contains m end lines, G has at least m/Δ independent end lines. Let G' be the graph obtained from G by deleting m/Δ independent end lines and let D denote the points of G' incident to the deleted lines.

Since K_ℓ contains no red $H = K(s_1, s_2, \ldots, s_t)$, we may sequentially pick red graphs $H_1, H_2, \ldots, H_{s_t - 1}$ such that each $H_i = K(s_1', s_2', \ldots, s_t')$, with $s_j' \leq s_j$ for all j, has the maximum number of points in $K_\ell - (H_1 \cup H_2 \cup \cdots \cup H_{i-1})$. Thus each point in $K_{\ell'} = K_\ell - (H_1 \cup H_2 \cup \cdots \cup H_{s_t - 1})$ has blue degree at least $s_t - 1$. For m/Δ sufficiently large ($m/\Delta \geq s^2$ works), the induction assumption implies that $K_{\ell'}$ contains a blue copy of G'.

If there is a blue matching between the points D of G' and the points of $K_\ell - G'$, then K_ℓ contains a blue G. Since this is not true, P. Hall's matching theorem implies the existence of a set L, $L \subset D$, such that its blue neighborhood in $K_\ell - G'$ has at most $|L| - 1$ points. Since the blue degree of each point in $K_{\ell'}$ is at least $s_t - 1$ in $K_\ell - G'$, we know in addition that $|L| \geq s_t$. Thus there are at least $\ell - (n-1)$ points in the red neighborhood of L in $K_\ell - G'$. From an inductive argument on t this red neighborhood contains a red $K(s_1, s_2, \ldots, s_{t-1})$ giving a red $H = K(s_1, s_2, \ldots, s_t)$ in K_ℓ. This contradiction completes this case and the proof of the proposition. □

Theorem 2. Let G be a connected graph such that $q(G) \leq p(G) + k$. Let $s_1 \leq s_2 \leq \cdots \leq s_t$, $s = \Sigma s_i$, and $\Delta = \Delta(G)$. Then

$$r(K(s_1, s_2, \ldots, s_t), G) = (t-1)(p(G)-1) + s_1 \quad \text{for}$$
$$n \geq 2\Delta(s^2 + (3s^2k + 2s^4\Delta)^{s-1})^2 + 3k(s^2 + (3s^2k + 2s^4\Delta)^{s-1} + 1.$$

The proof of this theorem is similar to that of Proposition 1. We do not present it, but indicate the parallel arguments. The lower bound is surely

$(t-1)(p(G)-1) + s_1$, since the graph $(t-1)K_{n-1} \cup K_{s_1-1}$ and its complement give an appropriate red-blue coloring of $K_{(t-1)(p(G)-1)+s_1-1}$. To show $(t-1)(p(G) = 1) + s_1$ is an upper bound, we induct on t. The induction step is completed by using both Lemma 1 and Proposition 1 and the ideas of the proof of Proposition 1. As in the proof of the proposition, one considers two cases. These cases are when G contains either a suspended path with $s^2 + (3s^2k + 2s^4\Delta)^{s-1}$ points or a set of $(s^2 + (3s^2k + 2s^4\Delta)^{s-1})\Delta$ points of degree 1.

Theorem 1 is essentially a restatement of Theorem 2.

4. OPEN QUESTIONS

There are several natural questions which should be considered. The main result is surely true for graphs G with more lines and smaller values of n than indicated in Theorems 1 and 2. Can these conditions as given be significantly improved? Examples show that the result fails when $t \geq 2$ and G is a star $K_{1,n}$, but it is likely that it will hold for $\Delta(G) = O(n)$. This should be considered. Also, if G has bounded degree and bounded line-density, is G an F-good graph for $p(G)$ large? Here line-density is defined as $\max q(H)/p(H)$, where $H \subset G$. Burr and Erdös have asked the parallel k-good question in [3].

There are many other similar problems to consider. A starting point would involve seeing which of the graphs in Theorem B (iii) - (vii) are F-good.

REFERENCES

1. S.A. Burr, Ramsey numbers involving graphs with long suspended paths. J. London Math. Soc. 24 (1981) 405-413.
2. S.A. Burr, Generalized ramsey theory for graphs - A survey. Springer, Lecture Notes Math. 406 (1974) 52-75.
3. S.A. Burr and P. Erdös, Generalizations of a ramsey-theoretic result of Chvátal. J. Graph Theory. 7 (1983) 39-51.
4. S.A. Burr and P. Erdös, Generalized ramsey numbers involving subdivision graphs, and related problems in graph theory. Ann. Discrete Math. 9 (1980) 37-42.
5. S.A. Burr, P. Erdös, R.J. Faudree, C.C. Rousseau, and R.H. Schelp, Ramsey numbers for the pair sparse graph-path or cycle. Trans. Amer. Math. Soc. 269 (1982) 501-512.
6. P. Erdös, R.J. Faudree, C.C. Rousseau, and R.H. Schelp, Multipartite graph-sparse graph ramsey numbers. Combinatorica (submitted).
7. S.A. Burr and J.W. Grossman, Ramsey numbers of graphs with long tails. Discrete Math. 41 (1982) 223-227.
8. V. Chvátal, Tree-complete graph ramsey numbers, J. Graph Theory 1 (1977) 93.
9. R.J. Faudree, C.C. Rousseau, and R.H. Schelp, Generalizations of a ramsey result of Chvátal. The Theory and Applications of Graphs (G. Chartrand, ed.) Wiley, New York (1981) 351-361.
10. R.J. Gould and M.S. Jacobson, On the ramsey number of trees versus graphs with large clique number. J. Graph Theory. 7 (1983) 71-78.

11. F. Harary, The foremost open problems in generalized ramsey theory. _Congressus Numerantium_ 15 (1975) 269-282.

12. F. Harary, _Graph Theory_. Addison-Wesley, Reading, Mass. (1969).

CONDITIONAL COLORABILITY IN GRAPHS

Frank Harary[1]

Department of Mathematics
University of Michigan
Ann Arbor, MI 48109

ABSTRACT. One of the most important invariants of a graph is its chromatic number and this interest has not been diminished by the computer-assisted proof of the Four Color Theorem. The chromatic number $\chi(G)$ has been generalized in several different directions. We propose a formulation which integrates these various approaches. This general concept, called conditional colorability, associates a prescribed graph theoretic property P with the parts of a partition of the vertex set V or the edge set E of the graph G. Then the conditional chromatic number (or the conditional chromatic index) of G with respect to P is the minimum number of parts in such a partition of V (or of E). The aspects of the chromatic number of G to be considered from the viewpoint of conditional colorability include the chromatic polynomial, critical graphs, the achromatic number, and associated questions in topological graph theory.

---

[1]Ulam Professor Mathematics, University of Colorado, Boulder, fall 1982; Research Professor of Electrical Engineering and Computer Science, Stevens Institute of Technology, Hoboken, NJ, spring 1984.

1. INTRODUCTION

Perhaps the two most intensively investigated invariants of a graph G are its chromatic number and its chromatic index. The notation and terminology of [11] is followed, except that a graph $G = (V,E)$ has *vertex set* V and *edge set* E. By definition, the *chromatic number* $\chi(G)$ is the minimum n such that there is an n-partition $V = \cup V_i$ in which no two adjacent vertices are in the same V_i (have the same color). Similarly, in the *chromatic index* $\chi'(G)$, $E = \cup E_j$ is partitioned and no two adjacent edges have the same color. (Regarding an edge as the set of its two vertices, two edges are adjacent when their intersection is a singleton.)

There are several other invariants of G which involve partitioning V or E. These include the arboricity $\Psi(G)$ discovered by Nash-Williams [22], the point arboricity $\rho(G)$ first studied by Chartrand, Kronk and Wall [6], the linear arboricity $\Xi(G)$ introduced in [13], the thickness $\theta(G)$ named by Tutte [26], the generalized chromatic number of Chartrand, Geller and Hedetniemi [5], and others. A natural formulation called "conditional colorability" is now stated which provides a unified viewpoint toward all such partitioning invariants. This was suggested by the recent paper [14] on conditional connectivity.

2. CONDITIONAL CHROMATIC NUMBER AND CHROMATIC INDEX

Let P be any graph theoretic property, e.g., bipartite, eulerian, hamiltonian, complete etc., and write $G \in P$

when G has property P. Then the conditional chromatic number $\chi(G:P)$ of graph G with respect to property P is the minimum n for which there exists an n-partition $V = \cup V_i$ such that for each i, the induced subgraph $\langle V_i \rangle$ is in P. Similarly, in defining the conditional chromatic index $\chi'(G:P)$, there is an n-partition $E = \cup E_j$ such that every edge-induced subgraph $\langle E_j \rangle \in P$. When P is the property that a graph F is forbidden to be induced by some vertices in the same V_i (or edges in the same E_j), it is convenient to write $P = -F$. Similarly when P forbids any member of a Family $\mathcal{F}$ of graphs, then $P = -\mathcal{F}$. Using this notation, the chromatic number of G is of course $\chi(G) = \chi(G: -K_2)$ and its chromatic index is $\chi'(G) = \chi'(G: -P_3)$ where P_3 is the path of order 3.

3. OLD AND NEW EXAMPLES OF CONDITIONAL COLORABILITY

We now discuss the conditional chromatic number $\chi(G:P)$ and index $\chi'(G:P)$ for nine graphical properties P, one at a time. We shall find that several but not all of these invariants have already been defined and studied in the literature.

1. $P = -K_2$.

As mentioned above, $\chi(G: -K_2)$ is the conventional chromatic number $\chi(G)$. Clearly the corresponding chromatic index is meaningless.

2. $P = -P_3$.

For the chromatic index, we have noted that $\chi'(G: -P_3) = \chi'(G)$. The corresponding chromatic number $\chi(G: -P_3)$ and in general $\chi(G: -P_n)$ is precisely the generalization of $\chi(G)$ proposed by Chartrand, Geller and

Hedetniemi [5]. However $\chi'(G: -P_n)$ is apparently new for $n \geq 4$.

3. $P = -K_3$.

For this conditional chromatic number, every planar graph G satisfies $\chi(G: -K_3) \leq 2$, as noted in Harary and Kainen [19]. The conditional chromatic index $\chi'(G: -K_3)$ does not appear to have been studied yet, nor have the larger complete graphs $\chi(G: -K_n)$ or $\chi'(G: -K_n)$ for $n \geq 4$.

4. $P = -\mathcal{C}$.

Here $\mathcal{C}$ is the family of all cycles $\{C_3, C_4, \ldots\}$. Chronologically, the chromatic index $\chi'(G: -\mathcal{C})$ was defined as the arboricity of G and was elegantly expressed by Nash-Williams [22] in terms of the induced subgraphs H of G as the maximum of the ceiling of the ratio $q(H)/(p(H) - 1)$ where p and q are the order and size (number of edges), respectively. More recently, the point arboricity was defined by Chartrand, Kronk and Wall [6] as $\rho(G) = \chi(G: -\mathcal{C})$.

A graph G has been called critical (actually, minimal) with respect to property P if $G \in P$ and for each edge $e \in E(G)$, $G-e \notin P$. It was shown in Bollobás and Harary [3] that point arboricity minimal graphs exist for both odd and even values of $\rho(G)$. The opposite phenomenon states that G is P-anticritical (actually, P-maximal) if $G \in P$ but for all $e \in E(\bar{G})$, $G + e \notin P$. Maximal graphs for several properties P were characterized by Harary and Thomassen [21].

5. $P = (G: -K_{1,n})$.

Of course forbidding the star is the same as the maximum degree condition, $\Delta(G) < n$. The conditional chromatic number $\chi(G:P)$ was studied by Cowen [7] for planar graphs G, but the index $\chi'(G:P)$ is new.

6. The property P for bipartite graphs is $P = -C_o$, the family of all odd cycles. The conditional chromatic index $\chi'(G: -C_o)$ was called the biparticity of G in Harary, Hsu and Miller [18]. This was denoted by $\beta(G)$ and it was shown that $\beta(G) = \lceil \log_2 \chi(G) \rceil$. The corresponding conditional chromatic number $\chi(G : -C_o)$ could be regarded as the bipartite version of point arboricity and has not yet been studied.

7. In a linear forest introduced in [13], each component is a path. When P is the linear forest property, one can write $P = -C, \Delta \leq 2$, the conjunction of two conditions. The linear arboricity $\Xi(G)$ is then just the conditional chromatic index, χ'(G: linear forest). The corresponding "linear point-arboricity" χ(G:linear forest) suggests itself immediately but has not yet been investigated. A conjecture for the exact value of $\Xi(G)$ for n-regular G, analogous to the result of Nash-Williams for $\Psi(G)$, has been settled by Akiyama, Exoo and Harary [1] for $n \leq 6$.

8. Let $H(\mathcal{F})$ be the family of graphs homeomorphic to some member of $\mathcal{F}$. Then by Kuratowski's Theorem, G is planar whenever $G \in P = -H(K_5, K_{3,3})$. The thickness $\theta(G)$ is precisely $\chi'(G:P)$. However the corresponding conditional chromatic number χ(G:planar), the "point thickness" of G has not yet been considered.

9. Now let D be a digraph and let $\vec{C}$ be the family of all directed cycles. Then the conditional chromatic number (which could be considered as the directed point-arboricity of a digraph) $\chi(D: -\vec{C})$ was used by Neumann-Lara [23] to derive a generalization of the following result of Erdős and Hajnal [9]:

In any graph G, there is an odd cycle of length $n > \chi(G) - 1$.

4. GENERALIZATIONS OF CHROMATIC NUMBERS

Other types of chromatic invariants have also been studied, which do not immediately express themselves in conditional terms. Stahl [25] defined $\chi_n(G)$ as the minimum number k of colors from which an n-set of colors can be assigned to each $v \in V$ so that whenever u and v are adjacent, their color sets are disjoint.

The line-distinguishing chromatic number $\chi_1(G)$ is the minimum number of colors which can be assigned to $V(G)$ so that for any two edges $e, e' \in E$, the 1-set or 2-set of colors on their vertices are distinct sets. It was shown by Frank, Harary and Plantholt [11] that the determination of $\chi_1(G)$ achieves the solution of a bin-packing problem. This gives another example of an application of a graph coloring concept to a type of problem in operations research in addition to the well-known use of $\chi'(G)$ for scheduling problems. The point-distinguishing chromatic index $\chi_o'(G)$ involves edge coloring and is defined similarly by Harary and Plantholt [20].

5. PROPOSED PROBLEM AREAS

The well-developed theories of the chromatic number and the chromatic index (see [4,10,12]) can now be extended to conditional colorability by investigating the following problem areas.

A. Bounds on the conditional chromatic index

The classic theorem of Vizing [27] proves that every graph G satisfies $\Delta \leq \chi' \leq \Delta + 1$. The implications of this basic result are intensively studied in the definitive book by Fiorini and Wilson [10] on edge colorings. What are the corresponding results concerning $\chi'(G:P)$ for the properties P of the preceding section, and other properties?

B. Minimal and Maximal Graphs

Minimal graphs with respect to the property that $\chi(G) = n$ were introduced by Dirac [8] who called such graphs "critical". Results on this topic are presented in [2, Chap. 5]. Maximal n-chromatic graphs are studied in [21]. What are the minimal and maximal graphs for the conditional chromatic numbers and indexes?

C. Uniquely Colorable Graphs

It was proved in Harary, Hedetniemi and Robinson [17] that for every $n > 3$, there is a uniquely n-colorable graph (where the χ color classes constitute a unique partition of V) which does not contain a complete subgraph K_n. In [2, p. 261], the definitive book on extremal graph theory by Bollobás, it is reported that a uniquely n-colorable graph can have arbitrarily large girth, and that this even holds for the conditional chromatic number when P is the property that a graph is "k-degenerate".

D. Chromatic Polynomials

In work not yet published, R. Bari defines and studies the "homomorphism polynomial" of a graph, and A.J. Schwenk also in an unpublished paper, introduces a new "unlabeled chromatic polynomial". These variations as well as the conventional chromatic polynomial surveyed by Read [24] can be investigated from the generalized viewpoint of conditional colorability.

E. Interpolation

The statement, $\chi(G) = n$, is equivalent to the fact that n is the minimum number for which there is a homomorphism of G onto K_n. In Harary, Hedetniemi and Prins [16], the corresponding maximum n was called the *achromatic number* of G and was denoted by $\psi(G)$. It was proved that for any graph G and for all n such that $\chi(G) < n < \psi(G)$, there is a homomorphism of G onto K_n. In a general approach to maximum versus minimum invariants for graphs [15], ψ was denoted by χ^+, and other interpolation results were obtained for different graph invariants. This question applies to each $\chi(G:P)$ and $\chi'(G:P)$, including the new invariants such as the vertex-thickness, $\chi(G:\text{planar})$, and the forbidden triangle chromatic index, $\chi'(G: -K_3)$.

REFERENCES

1. J. Akiyama, G. Exoo, F. Harary, Covering and packing in graphs IV: Cyclic and acyclic invariants. *Math. Slovaca* 80 (1980), 405-417.
2. B. Bollobás, *Extremal Graph Theory*. Academic Press, London (1978).
3. B. Bollobás and F. Harary, Point arboricity critical graphs exist. *J. London Math. Soc.* 12 (1975), 97-102.

4. J.A. Bondy and U.S.R. Murty, Graph Theory with Applications. Macmillan, London (1976).

5. G. Chartrand, D. Geller, S. Hedetniemi. A generalization of the chromatic number. Proc. Cambridge Philos. Soc. 64 (1968), 265-271.

6. G. Chartrand, H.V. Kronk, C.E. Wall, The point arboricity of a graph. Israel J. Math. 6 (1968), 169-175.

7. R. H. Cowen, Defective colorings of planar graphs, (to appear).

8. G.A. Dirac, A property of 4-chromatic graphs and some remarks on critical graphs. J. London Math. Soc. 27 (1952), 85-92.

9. P. Erdös and A. Hajnal, On chromatic numbers of graphs and set systems. Acta Math. Acad. Sci. Hungar. 17 (1966), 61-99.

10. S. Fiorini and R.J. Wilson, Edge-Colourings of Graphs. Pitman, London (1977).

11. O. Frank. F. Harary, M. Plantholt, The line-distinguishing chromatic number of a graph. Ars Combin. (to appear).

12. F. Harary, Graph Theory. Addison-Wesley, Reading (1969).

13. F. Harary, Covering and packing in graphs I. Ann. N.Y. Acad. Sci. 175 (1970), 198-205.

14. F. Harary, Conditional connectivity. Networks 13 (1983), 347-357.

15. F. Harary, Maximum versus minimum invariants for graphs. J. Graph Theory 7 (1983), 275-284.

16. F. Harary, S. Hedetniemi, G. Prins, An interpolation theorem for graphical homomorphisms. Portugal. Math. 26 (1967), 453-462.

17. F. Harary, S. Hedetniemi, R.W. Robinson, Uniquely colorable graphs. J. Combin. Theory 6 (1969), 264-270; 9 (1970, 221.
18. F. Harary, D. Hsu, Z. Miller, The biparticity of a graph. J. Graph Theory 1 (1977), 131-133.
19. F. Harary and P.C. Kainen, On triangular colorings of a planar graph. Bull. Calcutta Math. Soc. 69 (1977), 393-395.
20. F. Harary and M. Plantholt, The point-distinguishing chromatic index. This volume.
21. F. Harary and C. Thomassen, Anticritical graphs. Math. Proc. Cambridge Philos. Soc. 79 (1976), 11-18.
22. C.St.J.A. Nash-Williams, Edge-disjoint spanning trees of finite graphs. J. London Math. Soc. 36 (1961), 445-450.
23. V. Neumann-Lara, The dichromatic number of a digraph. J. Combin. Theory B33 (1982), 265-270.
24. R.C. Read, An introduction to chromatic polynomials. J. Combin. Theory 4 (1968), 52-71.
25. S. Stahl, n-tuple colorings and associated graphs. J. Combin. Theory 20(1976), 185-203.
26. W.J. Tutte, On the non-biplanar character of the complete 9-graph. Canad. Math. Bull. 6 (1963), 319-330.
27. V.G. Vizing, On an estimate of the chromatic class of a p-graph (Russian). Diskret. Analiz. 3 (1964), 25-30.

SOME IMPOSSIBILITY THEOREMS AND THE CONSENSUS PROBLEM FOR GRAPHS

Frank Harary[1]

Department of Mathematics
University of Michigan
Ann Arbor, MI 41809

F. R. McMorris[2]

Department of Mathematics
Bowling Green State University
Bowling Green, OH 43403

ABSTRACT. The famous Impossibility Theorem of Arrow is shown to state that for arc-colored multidigraphs, there is no realization of a certain system consisting of three axioms. This means that in any such structure, two of these conditions imply the negation of the third. The general notion of consensus for graphs is also discussed.

1. INTRODUCTION

Probably the best known and most interesting result in the extensive theory of social choice is the Impossibility Theorem of Kenneth Arrow. This theorem states that a

[1] Ulam Professor of Mathematics, University of Colo., Fall, 1982.

[2] Research partially supported by the Faculty Research Committee of Bowling Green State University

voting procedure, which satisfies three intuitively appealing conditions, cannot exist. In axiomatic terminology, the result shows that the system whose three axioms are called say (P), (I), and (D), without loss of generality, is inconsistent. In more conventional implication form this says that if (P) and (I) hold then (D) does not.

Since Arrow's 1951 monograph, there has been a plethora of papers and books written that were stimulated by this ingenious approach to group consensus. Our purpose in this expository note is to develop an approach to graph theoretic versions of this theorem, propose some open questions, and suggest what we feel to be a fruitful research program motivated by these considerations. Examples of other uses of graph theory in social choice can be found in [2, 6, 10].

2. A BRIEF REVIEW OF ARROW'S THEOREM

Let V be a finite set of alternatives, $v_1, v_2, \ldots, v_p$, to be ranked by each of k voters, $1, \ldots, k$. As is often the case, we require a ranking to be a weak order on V, i.e. a reflexive, transitive, complete relation, but not necessarily asymmetric (in which case it would be a complete order). Let W be the set of all weak orders of V and W^k the k-fold Cartesian product.

A *consensus function* is a map: $C : W^k \to W$, C for consensus, with the image of an element of W^k under C interpreted as the group consensus of the k voters. Much of the work in social choice theory has been devoted to the analysis of certain properties of various consensus functions.

For $\rho \in W^k$, let ρ^* be the relation defined by $x\rho^*y$ if and only if $x\rho y$ and not $y\rho x$. Thus in digraph terms, relation ρ^* is obtained from ρ by deleting all symmetric pairs of arcs of ρ; in other words ρ^* is the maximal asymmetric subrelation of ρ. Naturally $x\rho^*y$ can be thought of as stating that alternative x is strictly preferred to alternative y on the basis of ρ. Also note that the relation α defined by $x\alpha y$ if and only if $x\rho y$ and $y\rho x$ is an equivalence relation whose equivalence classes are considered to be the "indifference" classes of ρ. An element $P = (\rho_1,\ldots,\rho_k) \in W^k$ is called a *profile*. The profiles $P = (\rho_1,\ldots,\rho_k)$ and $P' = (\rho_1',\ldots,\rho_k')$ *agree* on $X \subset V$ if and only if $\rho_i \cap (X \times X) = \rho_i' \cap (X \times X)$ for all $i = 1,\ldots,k$. We denote the image of $P \in W^k$ under C by CP and abbreviate $(CP)^*$ by C^*P.

Three conditions listed by Arrow that might be imposed on a consensus function are the following:

(P) A consensus function C is *Pareto* if whenever $P \in W^k$ and $x,y \in V$ satisfy $x\rho_i^*y$ for $i = 1,\ldots,k$, then xC^*Py.

(I) A consensus function is *independent of irrelevant alternatives* if whenever the profiles P and P' agree on $X \subset V$, then CP and CP' agree on X.

(D) A consensus function is *nondictatorial* if for each j there is a profile P and elements $x,y \in V$ such that $x\rho_j^*y$ but not xC^*Py.

Condition (P), named after the Italian economist, Vilfredo Pareto [8], simply says that if every voter strictly prefers x over y, then so should society. Condition (I) requires that the consensus ranking of alternatives in X should depend only on the rankings of the voters for the alternatives in X. Condition (D) says that there should be no voter with the power that if she strictly prefers x to y then so should society regardless of how the other voters rank x versus y. Thus we see that, in a "democratic" society, all these conditions are reasonable. For more motivation and discussion, see [9] or [11].

We can now state the celebrated Impossibility Theorem of Arrow [1].

Theorem A. If $p \geq 3$, then there is no consensus function $C: W^k \to W$ that satisfies (P), (I), and (D).

Of course Theorem A simply says that if C is a consensus function satisfying (P) and (I), then there must be a "dictator" j, which means that if $x\rho_j^* y$ then xC^*Py for any $x,y \in V$.

3. A DIRECT ANALOGUE OF ARROW'S THEOREM

In this section we observe one of our main points: Theorem A can really be considered as a theorem about digraphs. If ρ is a weak order of V, then we can form an associated digraph in the obvious way by taking V as a vertex set and having an arc from x to y if and only if $x\rho y$. Call such a digraph a *weak order digraph*. Let $P = (\rho_1,\ldots,\rho_k)$ be a profile and D the superposition [3] of the k associated weak order digraphs. That is, D has vertex set V and xy is an arc with label i whenever xy is an arc in the ith weak order digraph. Suppose that each voter is assigned a color. Then D is k-arc-colored and each color induces a weak order subdigraph.

Let $\mathcal{P}_k$ be the set of all such multidigraphs obtained by superposing k weak order digraphs, each of which is arc-colored by a single color. For $x,y \in V$ let $s(xy)$, the *multiplicity of arc* xy, denote the number of arcs from x to y and $s(x,y) = s(xy) + s(yx)$.

A *consensus operation* C can now be considered to be a function $C: \mathcal{P}_k \to \mathcal{P}_{k+1}$ where, for each $D \in \mathcal{P}_k$, C(D) is obtained from D by superposing a new weak order digraph of a different color on to D. Since we have advertised an Arrow result, let us suppose that we must add these new arcs subject to the following constraints:

(P_1) If $D \in \mathcal{P}_k$ and $x,y \in V$ with $s(x,y) = k = s(xy)$, then in C(D) we must add an arc xy with the new color while not adding an arc yx with this color. (Note that this is the exact translation of the Pareto condition (P).)

(I_1) For $D,D' \in \mathcal{P}_k$ and $X \subset V$, if the submultidigraphs of D and D' induced by X are equal, then the submultidigraphs of C(D) and C(D') induced by X are equal. (Note that this is exactly the translation of condition (I).)

We can now state the following result which paraphrases Theorem A.

Theorem A'. If $C: \mathcal{P}_k \to \mathcal{P}_{k+1}$ for $p \geq 3$ is a consensus operation satisfying (P_1) and (I_1), then one color strictly dictates.

That is, there exists a color j, $1 \leq j \leq k$, such that whenever xy has color j in $D \in \mathcal{P}_k$ but yx does not, then we *must* add an arc xy in C(D) of the new color j while not adding an arc yx of this color.

4. VOTERS INDISTINGUISHABLE

For a profile $P = (\rho_1,\ldots,\rho_k)$, form the multidigraph defined by superposing the associated k weak order digraphs, but do not now distinguish the voters by colors. An example of this situation would be voting records which get mixed up; if the records that keep track of the weak orders of each voter were confused, then one would be unable to determine which weak order belonged to which voter. Let $\mathcal{A}_k$ be the family of all such multidigraphs.

Problem 1. Characterize the multidigraphs in $\mathcal{A}_k$.

This problem appears nontrivial and the following "easier" version is not completely settled.

Problem 2. Characterize multidigraphs that arise by superposing k transitive tournaments (complete orders).

A consensus operation is now considered to be a function $C: \mathcal{A}_k \to \mathcal{A}_{k+1}$ such that for $D \in \mathcal{A}_k$, $C(D)$ is required to superpose another weak order digraph on to D. Properties (P_1) and (I_1) are exactly analogous in this setting and we may make the following statement.

<u>Conjecture 1</u>. If $C: \mathcal{A}_k \to \mathcal{A}_{k+1}$ satisfies (P_1) and (I_1), then $C(D)$ is obtained from D by duplicating an induced weak order subdigraph, i.e., there is a "dictator" weak order subdigraph.

We feel that Problem 1 must be solved before attempting to prove or disprove Conjecture 1.

Other types of partially ordered sets that have been of interest in the social sciences are interval orders and semi-orders (see [9] for definitions).

Problem 3. Characterize those multidigraphs that arise by superposing interval and semi-order digraphs.

5. THE CONSENSUS PROBLEM FOR GRAPHS

Since most social choice theorems can really be considered results about digraphs of certain types, it seems natural to consider the notion of consensus for graphs in general. Here we assume all graphs have the same vertex set. The Consensus Problem for Graphs can be stated as follows: Given a collection of graphs $G_1, \ldots, G_k$, construct another graph G which captures, in some sense, the consensus or common agreement of all the G_i. Of course, this will be with respect to certain units of information or parameters of interest. For example, in the weak order context the important units of information are the order pairs of the relation. If graph G_i represents some societal structure as determined by sociologists; then notions of connectivity might be considered important when constructing a consensus structure of all the sociologists.

Formally, if $\mathcal{G}$ is a class of graphs satisfying certain conditions, then a *consensus function* is a map $C: \mathcal{G}^k \to \mathcal{G}$.

It might be feasible for various conditions to list, in the spirit of Arrow, certain reasonable properties that a consensus function should satisfy and then characterize those consensus functions, if any, that satisfy these conditions.

As examples, Mirkin [7] has proved Arrow type theorems, when $\mathcal{G}$ consists of transitive digraphs, while McMorris and Neumann [5] proved an Arrow theorem for directed trees. It would be particularly interesting to study consensus functions on undirected trees because of potential applications in numerical taxonomy.

When simply the existence of edges between vertices is deemed important, then there are several natural consensus functions. The most general of these we call a *Boolean consensus function* $B: \mathcal{G}^k \to \mathcal{G}$, where $\mathcal{G}$ is the class of all graphs and the edge set of $B(G_1,\ldots,G_k)$ is obtained by taking unions and intersections of the edge sets $E(G_i) = E_i$, $i = 1,\ldots,k$. For example, *unanimity rule* has edge set $E_1 \cap \cdots \cap E_k$, and *majority rule* has edge set $\cup(\bigcap_{I \in W} E_i)$ where the union is taken over all $W \subset \{1,\ldots,k\}$ such that $|W| > k/2$. For $\mathcal{H}$ a restricted class of graphs, there may be reasonable conditions that a consensus function $C: \mathcal{H}^k \to \mathcal{H}$ should satisfy. One would then seek a Boolean consensus function which gives a best approximation. This is left purposely vague since the conditions imposed would depend on the application at hand. However, we mention that the approximation might be with respect to the natural symmetric difference distance function. This metric is defined by $d(G_1,G_2) = |E_1 \wedge E_2|$ where $|E_1 \Delta E_2|$ is the symmetric difference of the edge sets E_1 and E_2; Margush [4] has characterized this distance function for graphs.

We hope to develop approaches to the problems and questions above and also to discuss impossibility theorems more generally in a later communication.

REFERENCES

1. K. J. Arrow, *Social Choice and Individual Values*. Wiley, New York (1951).
2. D. H. Blair and R. A. Pollak, A cyclic collective choise rule. *Econometrica* 50(1982) 931-943.
3. F. Harary and E. M. Palmer, *Graphical Enumeration*. Academic, New York (1973).
4. T. Margush, An Axiomatic Approach to Distances Between Certain Discrete Structures. Ph. D. thesis, Bowling Green State University (1980).
5. F. R. McMorris and D. Neumann, Consensus functions defined on trees. *Math. Social Sci.* 4(1983) 131-136.
6. E. Miller, Graphs and anonymous social welfare functions. *Int. Econ. Rev.* 23(1982) 609-622.
7. B. G. Mirkin, *Group Choice*. Winston, Washington (1979).
8. V. Pareto, *Cours d'économiè politique*. Rouge, Lausanne (1889).
9. F. S. Roberts, *Discrete Mathematical Models*. Prentice-Hall, Englewood Cliffs (1976).
10. F. Roberts, Graph theory and the social sciences. *Applications of Graph Theory* (R. J. Wilson and L. W. Beineke, eds.) Academic, New York (1979) 255-291.
11. A. K. Sen, *Collective Choice and Social Order*. Holden-Day, San Francisco (1970).

THE POINT-DISTINGUISHING CHROMATIC INDEX

Frank Harary
The University of Michigan
Ann Arbor, MI 48109

Michael Plantholt
Illinois State University
Normal, IL 61761

ABSTRACT. The point-distinguishing chromatic index of a graph $G = (V,E)$ is the smallest number of colors assignable to E so that no two distinct points are incident with the same color set of lines. The exact value of this new invariant is found for paths, cycles, complete graphs and n-cubes. In addition, bounds on this chromatic index are obtained for the complete bipartite graphs $K_{m,n}$ and these are tight when $m = n$. Finally, the minimum and maximum orders among all trees with given point-distinguishing chromatic index are obtained.

1. INTRODUCTION

Let f be a mapping of the set E of lines of G onto the k-set $N_k = \{1,2,\ldots,k\}$. The color induced by f of a point v with neighborhood $\{v_1,\ldots,v_r\}$ is $f(v) = \{f(vv_1), f(vv_2),\ldots,f(vv_r)\}$. Then f is a point-distinguishing coloring of G if for any two points u and w,

$f(u) \neq f(w)$. For any graph G which has no K_2 components and at most one isolated point, the <u>point-distinguishing chromatic index</u>, denoted $\chi_0(G)$, is the minimum number of colors used in any point-distinguishing coloring of G. Note that, we then have $\chi_0(G) \leq q$ since assigning each line a different color yields a point-distinguishing coloring.

If G contains a K_2 component or at least two isolated points, then obviously no point-distinguishing coloring exists. In this case we define for convenience $\chi_0(G)=\infty$.

Lemma 1. For any graph G and spanning subgraph H of G, $\chi_0(G) \leq \chi_0(H) + 1$.

<u>Proof</u>. If $\chi_0(H) = \infty$, the result is immediate. So suppose $\chi_0(H) = k$ and let f_H be a point-distinguishing k-coloring of the lines of H. Extend this to a (k+1)-coloring f_G of the lines of G by assigning all lines of $G - E(H)$ the color $k + 1$. Then f_G is a point-distinguishing $(k + 1)$-coloring of G because each point of G has a different subset of the colors in N_k incident with it. □

As in [2], we let p or p(G) denote the order of G.

Lemma 2. $\chi_0(G) \geq \log_2 p$.

<u>Proof</u>. The total number of distinct subsets of k colors is 2^k, so that if $\chi_0(G) = k$ then $2^k \geq p$. □

A point-distinguishing coloring of G is the dual of the line-distinguishing coloring which was introduced in [1]. The line-distinguishing chromatic number of G is written $\chi_1(G)$. However, in general $\chi_0(G)$ is not merely equal to $\chi_1(L(G))$ because in a point-distinguishing coloring, points may be assigned sets of any cardinality rather than just 1-sets and 2-sets. Based simply on this observation, one would expect the determination of $\chi_0(G)$ to be even more difficult in general than the determination of $\chi_1(G)$, and it is.

2. $\chi_0(G)$ FOR CERTAIN CLASSES OF GRAPHS

The line-distinguishing chromatic number of paths was found in [1] by identifying a line-distinguishing k-coloring of P_p with a trail of length $p-1$ in the universal relation K_k° constructed from K_k by adding a loop at each point:

$$\chi_1(P_p) = \min\left\{2\left\lceil\frac{1+\sqrt{8p-7}}{4}\right\rceil - 1,\ 2\left\lceil\sqrt{\frac{p-2}{2}}\right\rceil\right\}. \quad (1)$$

A similar method can be used to determine $\chi_0(P_p)$.

For $p \geq 3$, the point-distinguishing chromatic index of a path is given by:

$$\chi_0(P_p) = \min\left\{2\left\lceil\frac{1+\sqrt{8p-9}}{4}\right\rceil - 1,\ 2\left\lceil\sqrt{\frac{p-1}{2}}\right\rceil\right\}. \quad (2)$$

Proof of (2). Consider any point-distinguishing k-coloring f of the lines of P_p, and associate each of the k colors with a different point of K_k°. Then since f is point-distinguishing, traversing the lines of P_p will correspond to following a trail of length $p-2$ in K_k°. Also, because each endpoint of path P_p is incident with just one line, this trail in K_k° cannot include the loops at its (distinct) beginning and terminal points, as the endpoints of P_p essentially use up those loops. Consequently, $\chi_0(P_p)$ is the minimum k for which K_k° contains a trail of length p which begins and terminates with loops. For k odd, the longest such trail has length $k(k+1)/2 - 1$; for k even, the maximum length of such a trail is $k(k+1)/2 - (k/2 - 1)$. So $k(k+1)/2 - 1 \geq p$ if k is odd and $k(k+1)/2 - (k/2 - 1) \geq p$ if k is even. Solving for k yields $k \geq 2\lceil(1 + \sqrt{8p+9})/4\rceil - 1$ for odd k and $k \geq 2\lceil\sqrt{(p-1)/2}\rceil$ for even k. Then $\chi_0(P_p)$ is the smallest integer k satisfying these constraints. □

Because of the fact that the cycle is the only connected graph which is its own line graph, it transpires that

$$\chi_0(C_p) = \chi_1(C_p) , \tag{3}$$

which is given by:

$$\chi_0(C_p) = \min \{2\sqrt{p/2} , \quad 2\left\lceil\frac{1 + \sqrt{8p+1}}{4}\right\rceil - 1\} . \tag{4}$$

Proof of (4). Since $L(C_p) = C_p$ we must have $\chi_0(C_p) = \chi_1(C_p)$. The proof now follows from the determination of $\chi_1(C_p)$ in [1]. □

In view of Lemma 1, the next inequality follows immediately. If G is hamiltonian then

$$\chi_0(G) \leq \min \{2\lceil\sqrt{p/2}\rceil + 1, 2\left\lceil\frac{1 + \sqrt{8p+1}}{4}\right\rceil\}. \tag{5}$$

We turn next to the calculation of $\chi_0(K_p)$ for $p \geq 3$. By (5) we have an upper bound of approximately $\sqrt{2p}$. However, this can be greatly improved. First we need to introduce some new terminology. Let $P(k)$ denote the power set of $\{1,2,...,k\}$ and let G be a labelled graph with $V = \{v_1,...,v_p\}$. A <u>set assignment</u> for G is an assignment of one member S_i of $P(k)$

to each point v_i of G such that no two points are assigned the same set. A set assignment is realizable if there is a k-coloring f of the edges of G such that the set of colors which are assigned to lines incident with v_i is S_i. Obviously $\chi_0(G)$ equals the minimum number of colors used in any realizable set assignment of G. Now we can demonstrate the formula for the point distinguishing chromatic index of a complete graph with $p \geq 3$:

$$\chi_0(K_p) = \lceil \log_2 p \rceil + 1 \tag{6}$$

Proof of (6). Consider a realizable set assignment for K_p which assigns point v_i the color set S_i. Then for any i,j it follows that $S_i \cap S_j \neq \emptyset$ because S_i and S_j must both contain the color of the line $v_i v_j$. Thus for any realizable set assignment from k colors at most 2^{k-1} of the members of $P(k)$ can be assigned to points of K_n, since if a set S_i is assigned point v_i then the complement of S_i with respect to N_k cannot be assigned to any point. Therefore $2^{k-1} \geq p$ and so $k \geq \log_2 p + 1$.

To show that $\chi_0(K_p) = \lceil \log_2 p \rceil + 1$ it now suffices to supply a set assignment of K_p from $\lceil \log_2 p \rceil + 1$ colors and a corresponding coloring which realizes that set assignment. The cases $p = 3$ and $p = 4$ are verified in Figure 1.

So, assume $p \geq 5$ and let $k = 1 + \lceil \log_2 p \rceil$. Since $p \geq 5$ we have $p \geq k + 1$. Assign point v_i the set $\{1,i\}$ for $i = 1,\ldots,k$ and assign v_{k+1} the set N_k. Then stipulate that any remaining points be assigned unused subsets of N_k that contain color 1. We can do this without repeating any assignments since the number of

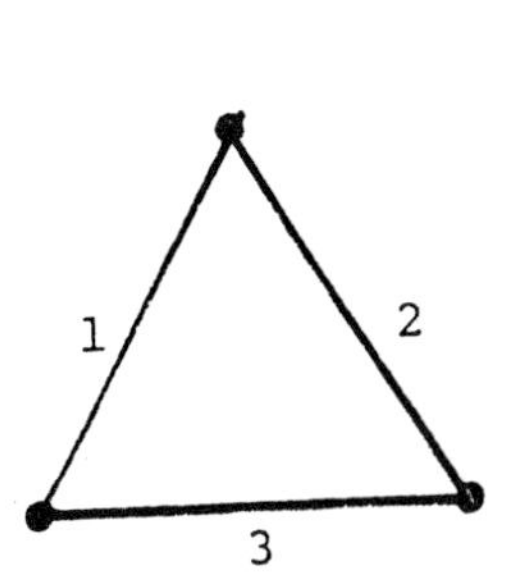

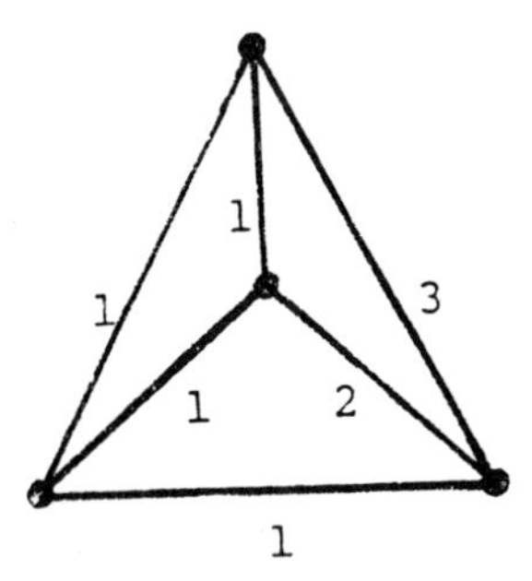

Figure 1. Optimal point-distinguishing colorings of K_3 and K_4.

subsets of N_k containing color 1 is $2^{k-1} \geq p$. Let S_i denote the set assigned point v_i for $i = 1, \ldots, p$. Color line $v_i v_j$ with the maximum numbered color in $S_i \cap S_j$. Obviously the set C_i of colors now assigned lines adjacent to v_i is a subset of S_i; it remains to show that $C_i = S_i$. It is clear that $C_1 = S_1 = \{1\}$. For $1 \leq i \leq k$ note that $v_1 v_i$ is assigned color 1 and $v_i v_{k+1}$ is assigned color i, so that $C_i = S_i$. Finally, for $i > k$ and $j \in S_i$, line $v_j v_i$ is assigned color j. Therefore $C_i = S_i$ for all i and the result follows. □

The determination of $\chi_0(K_{m,n})$ is even more difficult. Obviously we have the lower bound

$$\chi_0(K_{m,n}) \geq \lceil \log_2 (m+n) \rceil \tag{7}$$

by Lemma 2, but no exact formula is known. The following upper bound is sharp.

If $m \leq n$ and $m \geq \lceil \log_2 n \rceil + 1$, then

$$\chi_0(K_{m,n}) \leq \lceil \log_2 n \rceil + 2 . \tag{8}$$

<u>Proof of (8)</u>. Let $k = \lceil \log_2 n \rceil + 2$ and let the lines of $K_{m,n}$ be $u_1 v_j$ for $1 \le i \le m$, $1 \le j \le n$. Assign u_i the set $\{1,2,i+1\}$ for $i = 1,\ldots,k-1$ and complete the assignment of color sets to the u_i points with new subsets of N_k which contain $\{1,2\}$ as a subset. Assign v_1 the set $\{1\}$, V_2 the set $\{2\}$, and v_i the set $\{1,i\}$ for $i = 3,..,k$. Then complete the assignments of the v_i with subsets of $\{1,3,4,\ldots,k\}$ which contain $\{1\}$ as a subset. Now color line $u_i v_j$ with the maximum color in the intersection of the sets assigned u_i and v_j. By an argument similar to that used in the proof of (6), it can be verified that this coloring realizes the given set assignment. □

By combining the above lower and upper bounds the following result for n-regular complete bipartite graphs with $n \ge 2$ is obtained:

$$\lceil \log_2 n \rceil + 1 \le \chi_0(K_{n,n}) \le \lceil \log_2 n \rceil + 2. \qquad (9)$$

Both of the possible values given by (9) are realized. For example $\chi_0(K_{2,2}) = 3$ by (4), and Figure 2 verifies that $\chi_0(K_{5,5}) = 4$.

We turn next to the determination of $\chi_0(Q_n)$, where Q_n denotes the n-cube. Recall that the n-cubes can be defined recursively by $Q_1 = K_2$ and $Q_n = K_2 \times Q_{n-1}$. Since $Q_1 = K_2$, it follows that $\chi_0(Q_1) = \infty$, so assume $n \ge 2$ and let $\chi_0(Q_n) = k$. Now Q_n is n-regular, so that in any point-distinguishing k-coloring of Q_n each point will be assigned a set with cardinality between 1 and k. Therefore, because Q_n has cardinality 2^n we have $\sum_{i=1}^{k} \binom{k}{i} \ge 2^n$; consequently, $k \ge n+1$. We will now show that k is in fact equal to this lower bound, so that

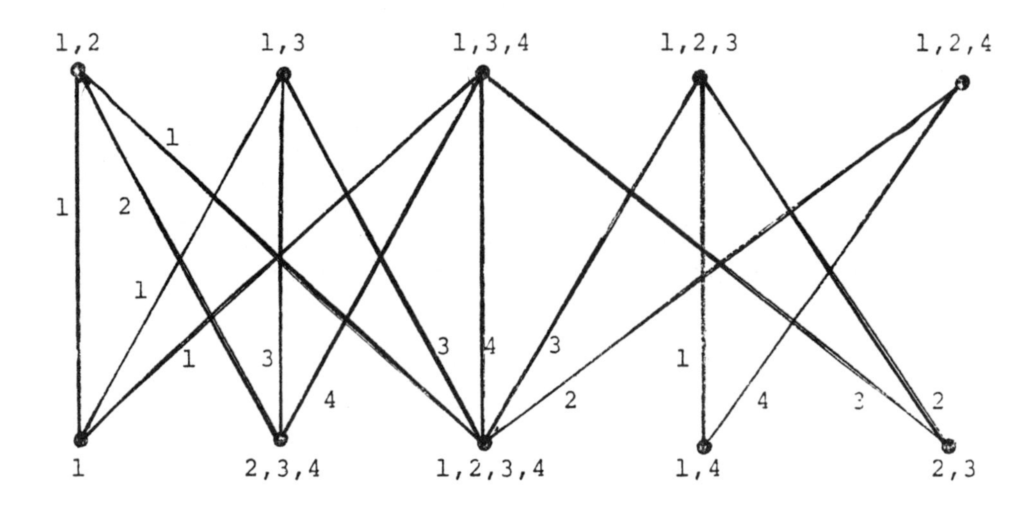

Figure 2. A point-distinguishing 4-coloring of $K_{5,5}$ and corresponding set assignment. Lines not shown may be assigned color 1 or color 2.

$$\chi_0(Q_n) = n + 1 \quad \text{for} \quad n \geq 2. \tag{10}$$

Proof of (10). By the preceding discussion, it suffices to construct a point-distinguishing (n+1)-coloring of Q_n for $n \geq 2$. Figure 3 shows a point-distinguishing 3-coloring of Q_2. Note that half the points of Q_2 are incident with the line assigned color 3, and there is a

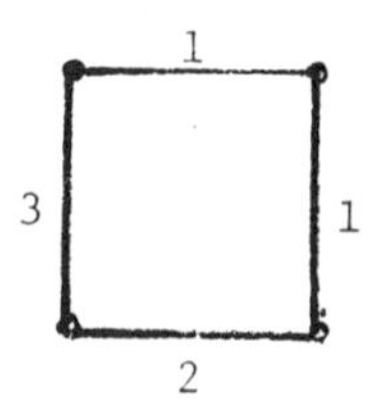

Figure 3. A point-distinguishing 3-coloring of Q_2.

set (of cardinality 1) of lines assigned color 1 which is not adjacent to any lines assigned color 3 but which covers the remaining two points of Q_2.

Now we assume that we have obtained a point-distinguishing n-coloring of Q_{n-1} using colors 1,...,n in such a way that 2^{n-2} points are incident with lines assigned color n, and there is a set of lines assigned color 1 which is not incident with any lines of color n, but which covers the remaining 2^{n-2} points of Q_{n-1}. Since $Q_n = K_2 \times Q_{n-1}$, it is convenient to think of Q_n as two distinct copies of Q_{n-1}, call them A and B, with each corresponding pair of points in the two copies joined by a line. Now color the lines of Q_n in the following manner.

Step 1. Color the lines of copy A of Q_{n-1} exactly as in the point-distinguishing n-coloring that is already known.

Step 2. Color the lines of copy B in the same way as the lines of copy A, but then make the following changes:

Recolor any line presently assigned color n with color n + 1. Now take all lines presently with color 1 which are not adjacent to any lines of color n + 1, and reassign them color n.

Step 3. Color the lines between copies A and B as follows: If a line is adjacent to a line of color n + 1, assign it color n + 1 also; otherwise, assign it color 1.

It is obvious that now 2^{n-1} points are incident with at least one line of color n + 1, and there is a set

of Q_n is, indeed, point-distinguishing. □

The colorings of Q_3 and Q_4 are shown in Figures 4 and 5.

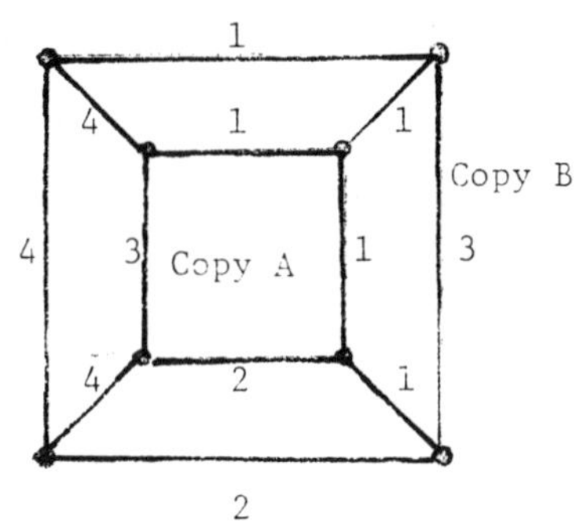

Figure 4. A point-distinguishing 4-coloring of Q_3.

COPY A

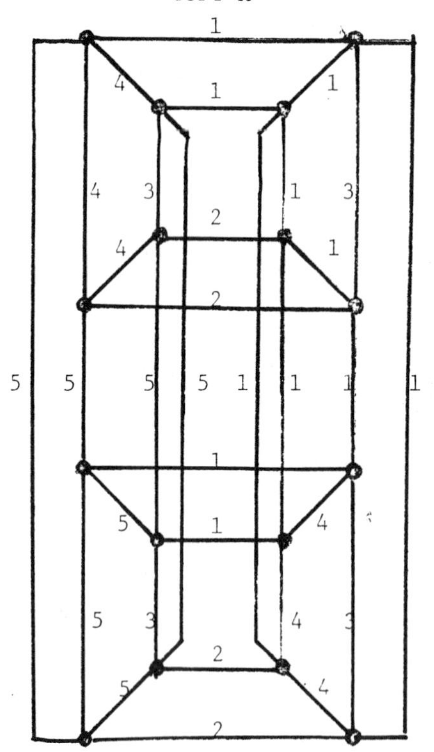

COPY B

Figure 5. A point-distinguishing 5-coloring of Q_4.

of lines with color 1 that is not adjacent to any line colored $n+1$ but which covers the remaining 2^{n-1} points. It remains to show that this $(n+1)$-coloring of Q_n is point-distinguishing.

To simplify the following discussion, we partition the points of Q_n into four sets of equal cardinality. Set W contains those points from copy A of Q_{n-1} which are incident to a line assigned color $n+1$; set X contains the remaining points from copy A. Set Y contains those points from copy B of Q_{n-1} which are incident to a line assigned color $n+1$; set Z contains the remaining points from copy B.

Now no two points in W have been assigned the same color set, because each inherited a different subset of $\{1,\ldots,n\}$ from the n-coloring of Q_{n-1}. The same argument shows that no two points in X have the same color set. Similar arguments hold for sets Y and Z since copy B receives a coloring similar to that of copy A.

It remains to verify only that no two points from distinct sets have been assigned the same color set. Each point of $W \cup Y$ contains color $n+1$ in its color set, so no point in $W \cup Y$ has the same color set as any point in X or Z. Also the color set of each point of W has color n and therefore cannot equal the color set of any point in X. Finally, no point in set X has color n in its color set. On the other hand, since in the point-distinguishing n-coloring of Q_{n-1} each point which is not incident with a line of color n must be incident with one of the special color 1 lines, it follows that each point of Z has n as a member of its set. Therefore, no color set of a point in X can equal that of a point in Z, so that the constructed coloring

3. EXTREMAL ORDERS OF TREES

Obviously, for $k \geq 2$, the minimum order among all trees with point-distinguishing chromatic index k is $k+1$, and the extremal tree in this case is the star $K_{1,k}$. We turn now to the opposite extremal problem, that of finding, for each $k \geq 2$, the maximum order among all trees with point-distinguishing chromatic index k; we denote this invariant by $M(k)$.

Lemma 3. For any $k \geq 2$, $M(k) \leq (k^2 + 3k - 4)/2$.

<u>Proof</u>. Let T be a tree with $\chi_0(T) = k$. Because there are only $\binom{k}{1} + \binom{k}{2}$ subsets of $\{1,2,\ldots,k\}$ with cardinality 1 or 2, T can have at most k points of degree 1 and at most $(k^2 + k)/2$ points with degree two or fewer. Also, since T is a tree, the average degree of its points is $2(p(T) - 1)/p(T)$. Therefore, T has at most $k-2$ points with degree greater than two and so can have a total of at most $(k - 2 + (k^2 + k)/2) = (k^2 + 3k - 4)/2$ points. □

We are now ready to show that the extremal value $M(k)$ is given by

$$\begin{aligned} M(k) &= (k^2 + 3k - 4)/2 \quad \text{for} \quad k \geq 2, \quad k \neq 4 \\ M(4) &= 11 \; . \end{aligned} \tag{11}$$

<u>Proof of (11)</u>. We consider first the case $k = 4$. Figure 6 displays a tree with order 11 whose point-distinguishing chromatic index is 4, so by Lemma 3 it follows that $M(4) = 11$ or 12.

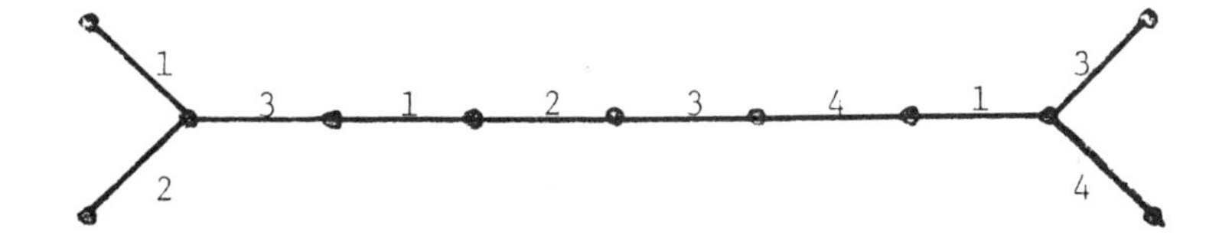

Figure 6. Extremal tree for $k = 4$.

Suppose that $M(4) = 12$ and that T is a tree such that $p(T) = 12$ and $\chi_0'(T) = 4$. Using the method of the proof of Lemma 3, it is easy to see that T must have exactly 4 points of degree 1, 6 points of degree 2, and 2 points of degree 3, and that in any point-distinguishing 4-coloring of T no lines assigned the same color can be adjacent. Consequently each color must be incident with one endpoint and exactly three points of degree 2 , and therefore with an even number (either 0 or 2) of points of degree three. We can therefore assume, without loss of generality, that colors 1, 2, and 3 are each incident with both the points of degree three. Thus both points of degree three have been assigned the color set $\{1,2,3\}$, a contradiction; consequently, $M(4) = 11$.

To show that $M(k) = (k^2 + 3k - 4)/2$ for $k \neq 4$, we need to construct a tree of appropriate order and display a point-distinguishing k-coloring. This can be done recursively, obtaining the extremal tree and coloring in k colors from the extremal graph and coloring in $k-2$ colors. We omit the proof, but give an indication of the construction by showing the extremal graph for $k = 5$ obtained from that for $k = 3$ in Figure 7, and the construction of the extremal graph for $k = 8$ from that for $k = 6$ in Figure 8.

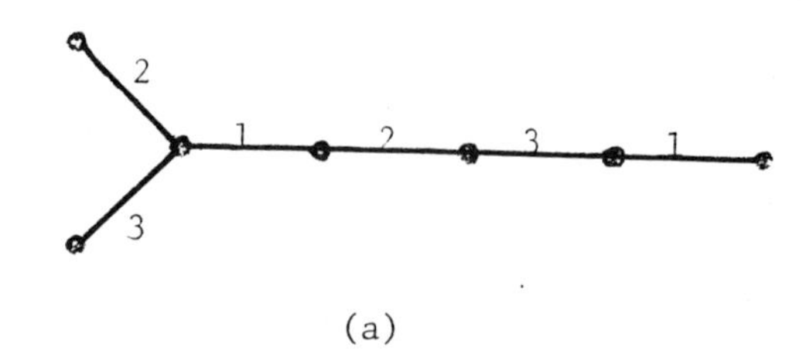

(a)

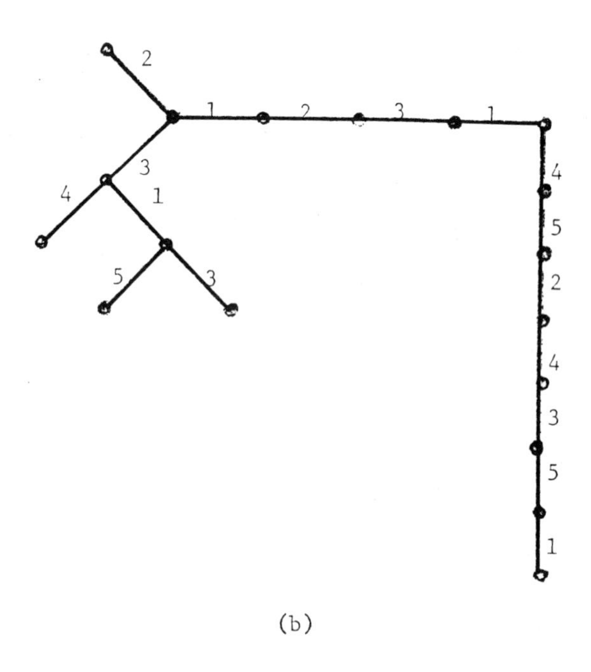

(b)

Figure 7. Extremal trees for $k = 3$ and $k = 5$.

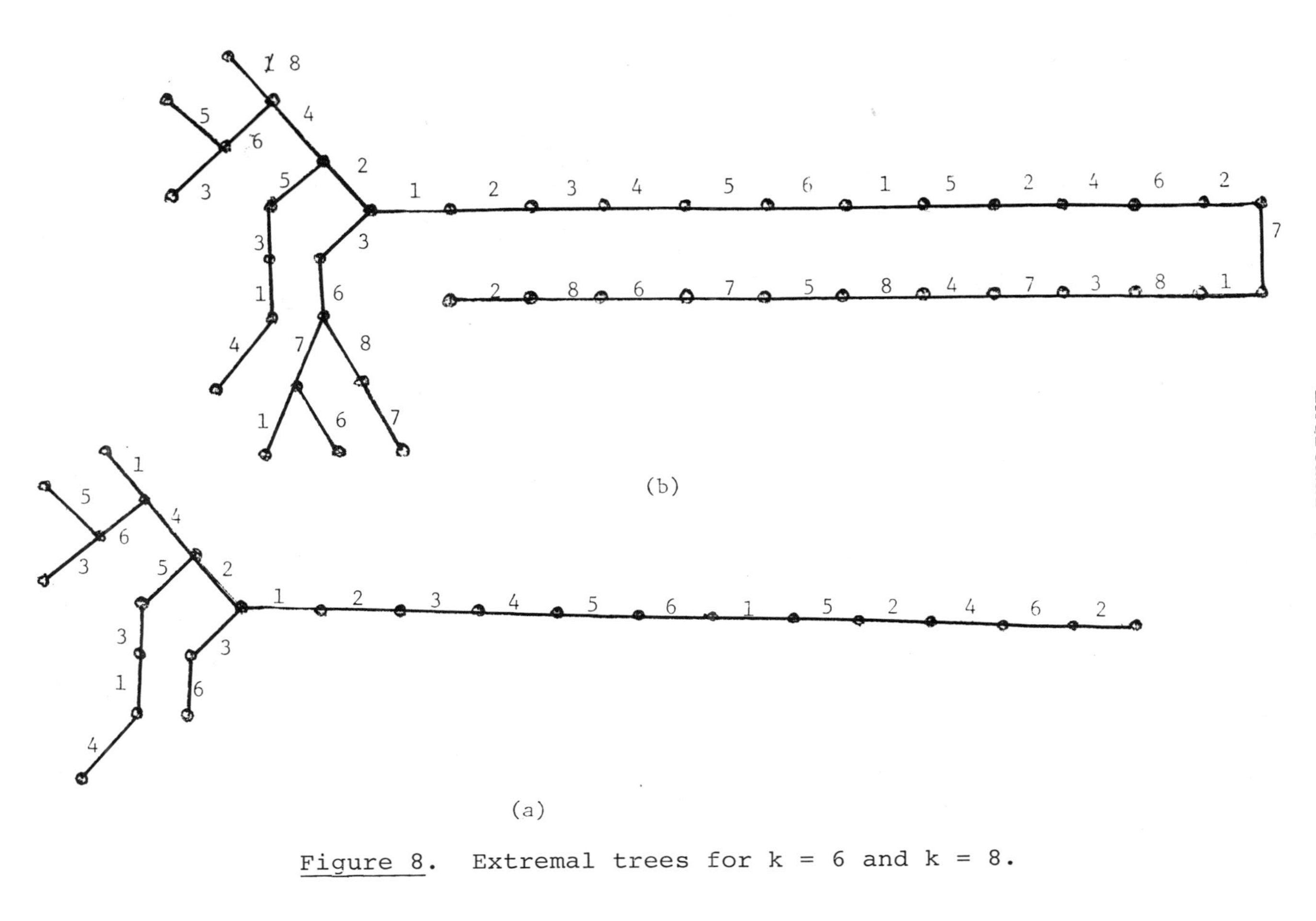

Figure 8. Extremal trees for k = 6 and k = 8.

4. UNSOLVED PROBLEMS

Determining $\chi_0(K_{n,n})$ exactly for arbitrary n is an unsolved problem. Another question is motivated by observing that $\chi_0(K_p)$ is equal to the intersection number of the complete graph, $\omega(K_p)$. For which graphs does $\chi_0(G)$ equal $\omega(G)$? This problem appears to be quite difficult. Another open question is to characterize the spanning subgraphs H of a graph G for which $\chi_0(G) = 1 + \chi_0(H)$. Finally when is a given set assignment of a graph realizable? Some necessary conditions related to systems of distinct representatives are available, but this problem also appears to be a very difficult one.

REFERENCES

1. O. Frank, F. Harary and M. Plantholt, The line-distinguishing chromatic number of a graph. Ars Combin. (to appear).

2. F. Harary, Graph Theory. Addison-Wesley, Reading (1969).

3. F. Harary and M. Plantholt, Graphs with the line distinguishing chromatic number equal to the usual one. Utilitas Math. (to appear).

THE MINIMAL BLOCKS OF DIAMETER TWO AND THREE

Frank Harary

Department of Mathematics
University of Michigan
Ann Arbor, MI 48109

Ralph Tindell

Department of Mathematics
Stevens Institute of Technology
Hoboken, NJ 07030

ABSTRACT. Our object is to provide a structural characterization of all blocks (nonseparable graphs) which are minimal (the removal of any line produces a subgraph that is not a block) and has diameter at most three, that is, we specify those low diameter graphs that are also minimal blocks.

1. INTRODUCTION

A graph will be taken as finite, non-empty, and having no loops or multiple lines as in [5], where all definitions regarding graphs not given here may be found. A connected graph G is a block if it has no cutpoints, and a block G is minimal if the removal of any line of G results in a subgraph that is not a block. Plummer [7] and Dirac [4] independently discovered and characterized the graphs in this important class. Recall that the diameter of a connected graph is the maximum distance between a pair of its points. We shall study those minimal blocks that also have low diameter d and provide a complete specification for $d \leq 3$. One reason for our interest is that tightly clustered or "dense" graphs play an essential role in the study of social networks, Barnes [1], including not just cliques (of diameter 1) but also the less densely clustered graphs of diameter 2 and 3.

We now summarize our results. It is clear that the only minimal blocks that have diameter 1 are K_2 and K_3. There are two types of minimal blocks of diameter two: the complete bipartite graphs $K_{2,p-2}$ and the graphs obtained from the pentagon by replacing each of two nonadjacent points by a "cloud" of nonadjacent points (see Figure 2). The exact specification of the minimal blocks having diameter three is considerably more complicated but is accomplished by first showing that every such graph may be obtained from one of the cycles C_6 or C_7 by a sequence of operations of the following type: Choose a path P in G of length $n \leq 3$ with its nonendpoints having degree 2 in G; now connect the endpoints u and v of G by a <u>new</u> path Q of length n' with $2 \leq n' \leq n$, that is, add Q to G so that $Q \cap G = \{u,v\}$. As defined below, this operation is called <u>handle reinforcement</u>.

It is worthwhile at this point to contrast the present problem with the work of Bollobás [2]; see also the expository article [3]. We consider here only graphs of diameter $d = 2$ or 3. Our object is to specify those graphs G satisfying (1) G has diameter d, (2) G is a block, and (3) if e is a line of G, then $G - e$ is not a block. On the other hand Bollobás [2] specifies those graphs G satisfying (1) G has diameter d, (2) G is a block, and (3') G has the fewest lines amongst all graphs satisfying conditions (1) and (2).

Notice that while we consider only minimal blocks, a graph satisfying (1), (2), and (3') need not be a minimal block as it may contain a line e which is not critical with respect to connectedness but rather is critical with respect to diameter. Moreoever most of the graphs specified in this paper are not such graphs, and we were unable to adapt Bollobás' methods to the present problem. Rather, our proofs proceed directly from Plummer's characterization of minimal blocks given in the next section.

2. BASIC RESULTS ON MINIMAL BLOCKS

The following characterization of minimal blocks, due to Plummer [7], is crucial to our arguments.

Plummer's Theorem. Let S denote the set of degree-2 points of a block G with at least 3 points. Then G is minimal if and only if either G is a cycle, or G-S is a disconnected forest such that for any component T of G-S and any cycle C in G, $C \cap T$ is either empty or connected.

We define a handle in a graph G to be a path H in G of length $\ell(H) \geq 2$, all of whose interior points lie in S, the set of degree-2 points of G. If G is a block and is not a cycle, then every point in S lies on a unique maximal handle. Our definition of a handle is a generalization of that given by Murty [6]. If distinct handles H_1 and H_2 in G have the same endpoints a and b, we say that H_2 reinforces H_1 and refer to $H_1 \cup H_2$ as a multiple handle in G. Also, in this situation, we say that G is obtained from its subgraph G_1 by <u>reinforcing</u> H_1 <u>by</u> H_2 provided $G = G_1 \cup H_2$ and $G_1 \cap H_2 = \{a,b\}$.

The following properties of minimal blocks were discovered independently by Dirac [3] and Plummer [7]:

P1. G has at least two vertices of degree 2.

P2. No handle of G has both endpoints in the same component of $G-S$.

P3. Either G is a triangle or G contains no triangles.

To this list we add one additional basic property:

P4. Every 4-cycle in G is a multiple handle in G.

<u>Proof</u>. Suppose C is a 4-cycle in G; then since $G-S$ is a forest, C contains at least one point of S. If all the points of C have degree 2 in the connected graph G, then $C = G$ and the conclusion follows. If either exactly one, or exactly two adjacent points of C lie in S, then P2 is violated. Thus C has exactly two points of degree 2 in G and they are not adjacent in C. Hence C is the union of two length-2 handles with common endpoints. □

2. THE RECURSIVE CHARACTERIZATION

We give in Theorem 2 a recursive characterization of minimal blocks of diameter 2 and 3 whose application in the next section leads to a complete structural specification. The primary tool needed in the proof of Theorem 2 is the existence of multiple handles. We first establish a preliminary technical result.

Lemma 1. Let G be a minimal block of diameter 2 or 3 containing no multiple handles. If H_1 and H_1' are maximal handles in G with a common endpoint w, the remaining endpoints lie in different components of $G-S$.

<u>Proof</u>. We prove Lemma 1 by contradiction. We assume that the endpoints u_1 of H_1 and u_2 of H_2 that are distinct from w lie in the same component tree T of $G-S$, and show that there is a cycle C in G whose intersection with T is not connected. Since G is a block and is not a cycle (which <u>is</u> a multiple handle), the endpoints of maximal handles lie in $G-S$.

Since G contains no multiple handles, $u_1 \neq u_1'$. Let P_1 and P_1' be disjoint paths in T conecting u_1 and u_1', respectively, to endpoints of T; if either point is an endpoint, the corresponding path is just the point itself. Since the endpoints of T have degree 1 in T but have degree 3 or more in G, each is the endpoint of at least two maximal handles. Thus the paths P_1 and P_1' contain points connected by maximal handles to points of $G-S$ other than w. Let u be the first such point encountered in traversing P from u_1, let H_1 be a maximal handle connecting u to $v \neq w$, and let Q be the portion of P_1 from u_1 to u. Define u', H_1', v', and Q' similarly.

If $v = v'$, then a cycle C violating Plummer's Theorem is obtained by setting

$$C = H_1 \cup Q \cup H_2 \cup H_2' \cup Q' \cup H_1' .$$

If v is joined by a line e to v', adding e to the above union provides the desired cycle. We may thus assume that v and v' are distinct and nonadjacent. Since v has degree at least 3 in G, there are two points adjacent to v but not on H_2. At least one of these must have distance at least two from w, else G contains either a 3-cycle or a 4-cycle; the former is forbidden by P3, the latter by the lack of multiple handles and P4. This argument also prevails for v', so let y and y' be points of distance at least 2 from w, y joined by a line e to v and y' by line e' to v'. By P2, neither y nor y' is in T.

Let P be a shortest path in G from y to y'. Since its length $\ell(P) \leq 3$, P contains at most two points from $Q \cup Q'$. If it contains none, then

$$C = H_1 \cup Q \cup H_2 \cup \{e\} \cup P \cup \{e'\} \cup H_2' \cup Q' \cup H_1'$$

is a cycle intersecting T in the disconnected set $Q \cup Q'$, as desired. If P contains exactly one point z of $Q \cup Q'$, then z is adjacent to one of the end-points of P, say y. Since y is not in T and z is in $T \subset G-S$, y has degree 2 in G, hence is the interior point of a maximal handle from v to z. Since u was the first point of Q connected by a maximal handle to a point other than w, z must be on Q'. If Q_1' is the segment of Q' from u_1' to z, then

$$C = H_1 \cup Q \cup H_2 \cup H_3 \cup Q_1' \cup H_1'$$

is a cycle violating the condition of Plummer's Theorem.

The only remaining case is when P contains two points z, z' of $Q \cup Q'$. We may assume z adjacent to

y and z' to y'. As before, z must be in Q' and z' in Q; also y is the interior point of a maximal handle H_3 from v to z, and y' similarly determines a maximal handle H_3' from v' to z'. If we let Q_1' be the segment of Q' from z to u' and let Q_1 be the segment of Q from z' to u, then

$$C = Q_1 \cup H_2 \cup H_3 \cup Q_1' \cup H_2' \cup H_2'$$

is a cycle, violating Plummer's Theorem. □

Theorem 1. If G is a minimal block of diameter 2 or 3, then G contains a multiple handle.

Proof. We will assume that G contains no multiple handles and derive a contradiction. For the moment, suppose G-S contains only points. Then any path in G connecting two points of G-S must have length at least 2, else G-S contains a line. Because shortest paths in G have length at most 3, any shortest path between points G-S must be a maximal handle--only its endpoints may be in G-S. Now, assuming that G contains no multiple handles, G is not a cycle, so G-S is nonempty. In fact, there must be at least 4 points in G-S: we assumed the components of G-S to be points, and they must be of degree 3 or more; thus each is connected by a handle to 3 distinct points of G-S. This implies the existence of two disjoint maximal handles, and one easily sees that the distance from an interior point of one to an interior point of the other must then be at least 4, a contradiction.

To complete the proof of Theorem 1, we need only show that if a minimal block G of diameter 2 or 3 contains no multiple handles, then G-S contains no lines. Suppose to the contrary that $u \neq u'$ are endpoints of a component tree T of G-S. Let H_1 and H_2 be maximal handles in G connecting u to points w_1, w_2 and let H_1' and H_2' be maximal handles connecting u' to points w_1' and w_2'. Then by Lemma 1, w_1 and w_2 lie in different components of G-S, as do w_1' and w_2'. We also note that the distance from any of the points w_1, w_2, w_1', w_2' to any point of T is at least 2--they all lie in other components of G-S.

Now choose paths P_1 and P_2 with the following properties: (1) Each path connects a point of $\{w_1, w_2\}$ to a point of $\{w_1', w_2'\}$; (2) P_1 and P_2 have no common endpoint; and (3) $\ell(P_1) + \ell(P_2)$ is as small as possible subject to conditions (1) and (2). By relabeling if necessary, we may assume P_i connects w_i to w_i', $i = 1,2$. Since G has diameter at most 3, we may assume $\ell(P_i) \leq 3$, $i = 1,2$.

We next show that P_1 and P_2 are disjoint. First notice that $\ell(P_1)$, $\ell(P_2) \leq 3$ and condition (2) imply that P_1 and P_2 have at most one line in common. Thus, if y is a point common to both paths, y has degree at least 3 in G. Thus y cannot be adjacent to both w_1 and w_2 as they lie in different components of G-S; similarly y is not adjacent to both w_1' and w_2'. This implies $\ell(P_1) = \ell(P_2) = 3$. Also, y must be adjacent to an endpoint of each of the paths; without loss of generality, we may assume y adjacent to w_1 and to w_2'. But this means there is a length-2 path

P_1' from w_1 to w_2' and letting P_2' be any path from w_2 to w_1' of length 3 or less, we obtain a violation of (iii).

Because P_1 and P_2 are disjoint, and neither intersects T, we may construct a cycle intersecting T in the set $\{u,u'\}$ by setting

$$C = H_1 \cup P_1 \cup H_1' \cup H_2' \cup P_2 \cup H_2 .$$

As $C \cap T = \{u,u'\}$ is disconnected, the conclusion of Plummer's Theorem is violated, a contradiction which completes the proof of Theorem 1. □

We now use this result to deduce the following recursive characterization.

Theorem 2. Let G be a minimal block of diameter $d = 2$ or $d = 3$. Then G may be obtained from a cycle of diameter d by successive reinforcement of handles of length $k \leq d$ by handles of length $j \leq k$.

Theorem 2 is an immediate corollary (by an inductive argument) of the following lemma and Theorem 1. □

Lemma 2. Let G be obtained from G_1 by reinforcing a handle H_1 with H_2 such that $\ell(H_2) \leq \ell(H_1) \leq d$, where $d = 2$ or $d = 3$.

(1) If G_1 is a minimal block of diameter d, then so is G.

(2) If G is not a cycle and is a minimal block of diameter d, the same holds for G_1.

Proof of (1). That G is a minimal block follows

directly from the definition. If $\ell(H_1) = \ell(H_2)$, then clearly the diameter of G is that of G_1; since this must be the case if $d = 2$, we may assume $d = 3$, and that $\ell(H_2) = 2$, $\ell(H_1) = 3$. Suppose the diameter of G is 2; then, because H_1 is a length 3 handle, every vertex of G not on H_1 must be adjacent to both endpoints of H_1. As this implies that the diameter of G_1 is 2, we have a contradiction.

Proof of (2). Again it is a straightforward matter to show that G_1 is a minimal block. We must show that the removal of the handle H_2 (exclusive of endpoints) from G does not increase the diameter. If $\ell(H_1) = \ell(H_2)$, this is obvious, and this observation takes care of the case $d = 2$. Hence we may assume $\ell(H_2) = 2$ and $\ell(H_1) = 3$, and we must show that the diameter of G_1 is 3. We need only consider those pairs of points in G_1 for which a shortest connecting path in G contains H_2. We notice that such a path must connect an endpoint u' of H_2 to a point v in G_1 which is adjacent to the other endpoint u of H_2. We must show there is a length-3 path in G_1 from u' to v. If v is in H_1 this is clear, so we assume v to lie in $G - (H_1 \cup H_2)$. Since G_1 is a block, there is a point $w \neq u$ with w adjacent to v. By property P3, w is not adjacent to u. Since there must be a path of length at most 3 from w to the interior point of H_1 adjacent to u', there must be a path of length at most 2 from w to u', and we are done. □

Remark. The arguments given for part 2 of Lemma 2 are peculiar to the case where a handle of length d is being reinforced by one of length $d-1$ or d. For $d \leq 3$, of course, this was sufficient. To see the special nature of the above argument, and to see that our strategy for the structural analysis of diameter-d minimal blocks fails for $d \geq 4$, one has only to consider the diameter-4 minimal block of Figure 1.

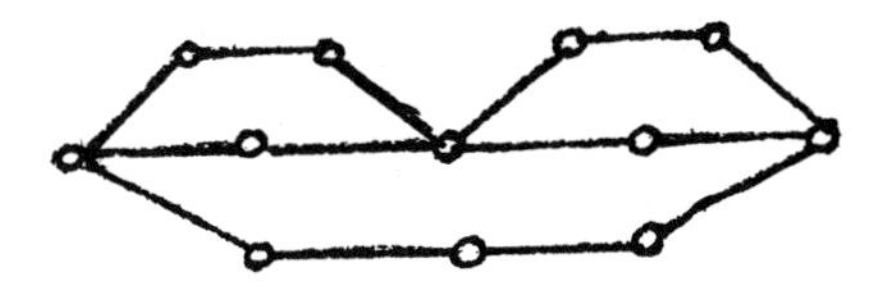

Figure 1. A minimal block on which we cannot get a handle.

3. THE STRUCTURE OF MINIMAL BLOCKS OF SMALL DIAMETER

In this section we indicate how the application of Theorem 2 leads to a complete specification of the minimal blocks of diameter 2 and 3.

First consider the case of diameter 2. According to Theorem 2, all minimal blocks of diameter 2 are obtained from one of the two cycles of diameter 2, C_4 or C_5, by successive reinforcement of length-2 handles by length-2 handles. If a single length-2 handle of C_4 is reinforced, the result is the complete bipartite graph $K_{2,3}$. Now, any further handle reinforcement will simply enlarge this system of multiple handles, and the general class of minimal blocks arising from C_4 is the class of complete bipartite graphs $K_{2,p-2}$ with $p \geq 4$.

Now consider the result $C(2,1)$ of reinforcing a length-2 handle H_1 of C_5 by a length-2 handle H_2, and suppose the endpoints of H_1 are a and b. If we wish to reinforce a handle of $C(2,1)$, we must reinforce either H_1 or a handle H_1' of C_5 sharing exactly one endpoint, say a, with H_1 and whose remaining endpoint c lies on neither H_1 nor H_2. If, after some number of reinforcements of H_1, the handle H_1' is reinforced, all remaining reinforcements may be viewed as reinforcements of either H_1 or H_1'. The graphs resulting from this procedure are the graphs $C(m,n)$ in which H_1 has been reinforced by m-1 handles and H_1' has been reinforced by n-1 handles, with $m, n \geq 1$. Using Dirac's terminology, the graph $C(m,n)$ may be described as the graph obtained by replacing two non-adjacent points of C_5 with "clouds" of m and n independent points, respectively. The graph $C(2,3)$ is depicted in Figure 2.

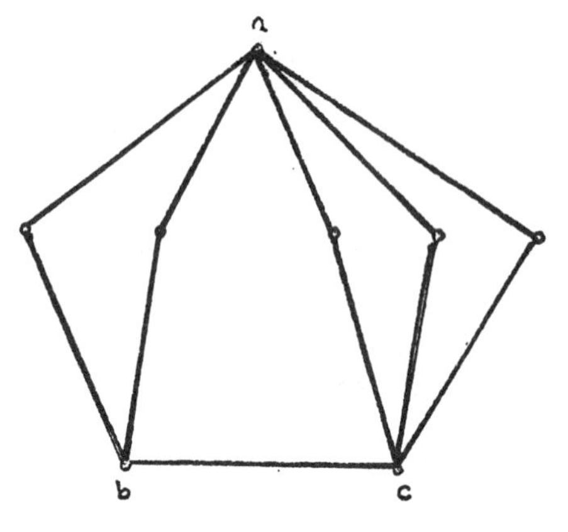

Figure 2. C(2,3)

Theorem 3. The minimal blocks of diameter 2 are the graphs $K_{2,p-2}$ with $p \geq 4$, and the graphs $C(m,n)$ with $m, n \geq 1$.

We now consider the minimal blocks of diameter 3. For this purpose, we define two classes of graphs. Let

$j, j', \ell, \ell', r, s, t, t', n$, and n' be nonnegative integers, and let $\underline{k}, \underline{k}', \underline{m}, \underline{m}'$ be positive integer vectors of length j, j', ℓ, ℓ' respectively, We define a graph $G(r,s,t,n;j,\underline{k};\ell,\underline{m})$ as follows: beginning with three points u, v, w, connect u to w by r length-2 handles; connect w to v by s length-2 handles; connect u to v by t length-2 handles and n length-3 handles; connect v by lines to new points $x_1, \ldots, x_j$ and, for each $i \leq j$, connect x_i to u by k_i length-2 handles; finally, connect u by lines to new points $y_1, \ldots, y_\ell$ and, for each $i \leq \ell$, connect y_i to v by m_i length-2 handles. If $r = s = 0$, it is understood that the point w is removed from G. The graph $G(2,2,2,2;2,(2,2);2,(2,2))$ is illustrated in Figure 3.

We now define the class of graphs $H(j,\underline{k};\ell,\underline{m};\ n,t,t',n';\ j',\underline{k}';\ell';\ m')$. Begin with three

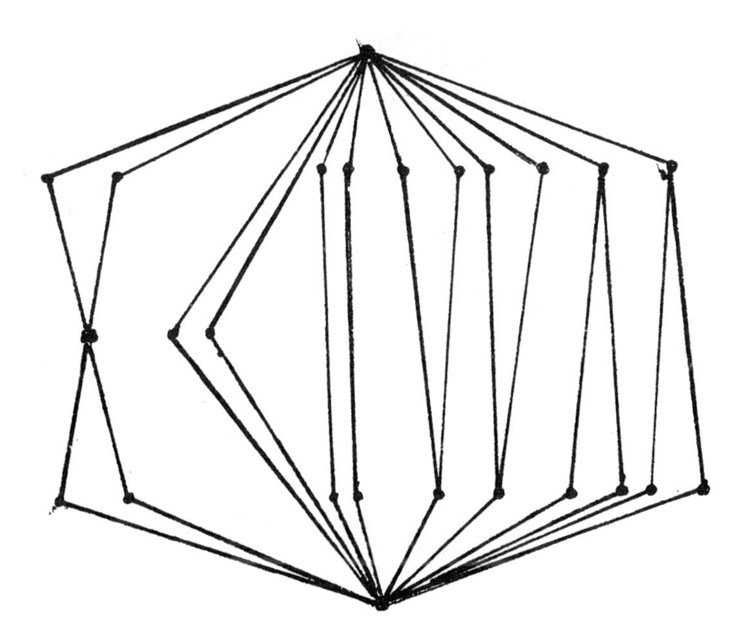

Figure 3. $G(2,2,2,2;2,(2,2);2,(2,2))$

points u, v, v' and join v and v' by a line; connect u to v by t length-2 handles and n length-3 handles; connect u to v' by t' length-2 handles and n' length-3 handles; connect v to new points $x_i, \ldots, x_j$ and, for each $i \leq j$, connect x_i to u by k_i length-2 handles; connect v' by edges to new points $x_1', \ldots, x_j'$ and for each $i \leq j'$, connect x_j' to u by k_1' length-2 handles; join u by lines to new points $y_1, \ldots, y_\ell, y_1', \ldots, y_\ell'$, for each $i \leq \ell$, connect y_i by m_i length-2 handles to v; and, for each $i \leq \ell'$, connect y_i' to v' by k_i' length-2 handles to v'. The graph $H(2,(2,2);2,(2,2);2,2,2,2;2,(2,2);2,(2,2))$ is illustrated in Figure 4.

We now consider the result of applying Theorem 2 to the construction of the minimal blocks of diameter 3.

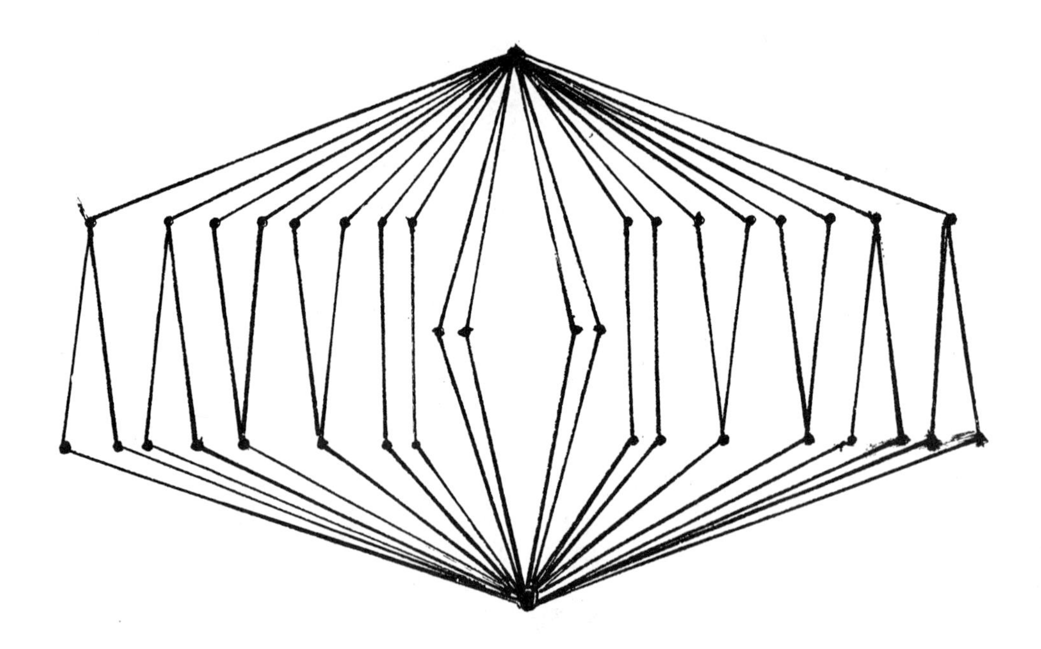

Figure 4. $H(2,(2,2);2,(2,2);2,2,2,2;2,(2,2);2,(2,2))$

If we begin with C_6 and reinforce only handles of length 2, the result is either the graph obtained by replacing three mutually nonadjacent points with clouds of independent points, or the graph in which the same is done for a pair of diagonally opposite points. The latter class of graphs clearly consists of the graphs $G(0,0,0,0;1,k_1;1,m_1)$: with $k_1, m_1 \geq 1$, and the former class consists of the graphs $C(r,s,t) = G(r,s,t,0;0;0)$ with $r,s,t \geq 1$. The graph $C(2,2,2)$ is illustrated in Figure 5.

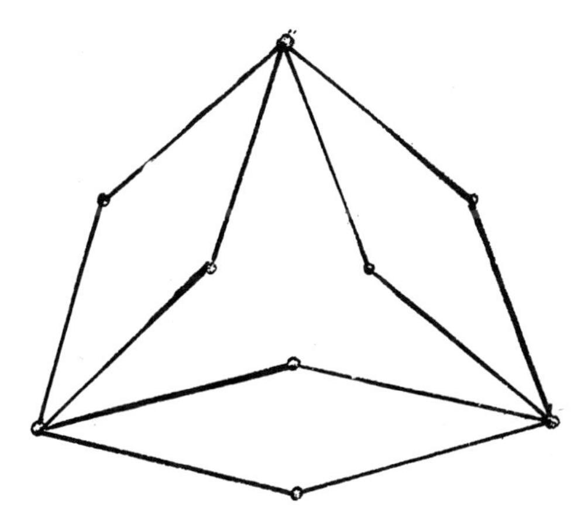

Figure 5. C(2,2,2)

If, beginning with C_6, our first step is to reinforce a length-3 handle H with endpoints u and v, subsequent steps must either be further reinforcement of H, or reinforcement of a length-2 subhandle of some length-3 handle. In the latter situation, there are only two possibilities, depending on which endpoint of the length-3 handle is shared by the subhandle. The graphs arising in this case are the $G(0,0,t,n;j,\underline{k};\ell,\underline{m})$ with $n + j + \ell \geq 2$. We have now exhausted the diameter-3 minimal blocks constructed from C_6.

We now apply the algorithm implicit in Theorem 2 to C_7. If we only reinforce length-2 handles, the resulting class of graphs will consist of the graphs $G(r,s,0,0;1,k_1;0)$ with $r,s,k_1 \geq 1$. If a length-3 handle of C_7 is reinforced and the remaining length-4 handle of C_7 is divided into two length-2 handles, which are then reinforced, the resulting graphs will be of the form $G(r,s,t,n;0;0)$ with $r,s,n \geq 1$. The only new reinforcements possible on graphs of the latter type involve length-2 subhandles of length-3 handles, and the general class of graphs for this case consists of the graphs $G(r,s,t,n;j,\underline{k};\ell,\underline{m})$ with $r,s,n+j+\ell \geq 1$. Note that this class contains the graphs $G(r,s,0,0;1,k_1;0)$ considered above.

Suppose now that two different length-3 handles of C_7 are reinforced. Then the two original handles will have one endpoint, designated as u, in common, and remaining endpoints v,v' distinct. If we reinforce the two length-3 handles with as many length-2 and length-3 handles as desired, the resulting class consists of the graphs of the form $H(0;0;n,t,t',n';0;0)$ with $n,n' \geq 1$. The only further possible reinforcements are of length-2 sub-handles of length-3 handles, and the resulting graphs are the graphs $H(j,\underline{k};\ell,\underline{m};n,t,t',n';j',\underline{k}';\ell',\underline{m}')$ with $j+\ell+n \geq 1$ and $j'+\ell'+n' \geq 1$. We have also exhausted the class of diameter-3 minimal blocks. These observations are summarized in the next statement.

Theorem 4. The following graphs are the minimal blocks of diameter 3:

(1) $C(r,s,t) = G(r,s,t,0;0;0)$ with $r,s,t \geq 1$.

(2) $G(r,s,t,n;j,\underline{k};\ell,\underline{m})$ with $r = 0$ iff $s = 0$, and $r+n+j+\ell \geq 2$.

(3) $H(j,\underline{k};\ell,\underline{m};n,t,t',n';j',\underline{k}';\ell',\underline{m}')$ with $j+\ell+n \geq 1$ and $j'+n'+\ell' \geq 1$.

4. THE EDGE MINIMUM MINIMAL BLOCKS OF LOW DIAMETER

As we now have a complete specification of the minimal blocks of diameter 2 and 3, we are in a position to determine, among the diameter 2 and 3 minimal blocks on p points, those for which the number q of lines is minimum. For diameter 2, we see that $K_{2,p-2}$ has $2p-4$ lines whereas $C(m,n)$ has $2p-5$ lines. Thus the graphs $C(m,n)$ are the "minimum" minimal blocks of diameter 2; see Figure 2.

One may similarly compare the number of lines in the various types of diameter-3 minimal blocks on p points, and determine those with a minimum number of lines. The number of lines in each "minimum" minimal block of diameter-3 is found to be $[3(p-2)/2]$, and this value is realized precisely by the graphs $G(0,0,0,n;0;0)$ with $n \geq 2$; $H(0;0;n,0,0,n';0;0)$ with $n \geq 1$ and $n' \geq 1$; and the graphs obtained from the latter by reinforcing a single length-2 handle by another of the same length; see Figure 6.

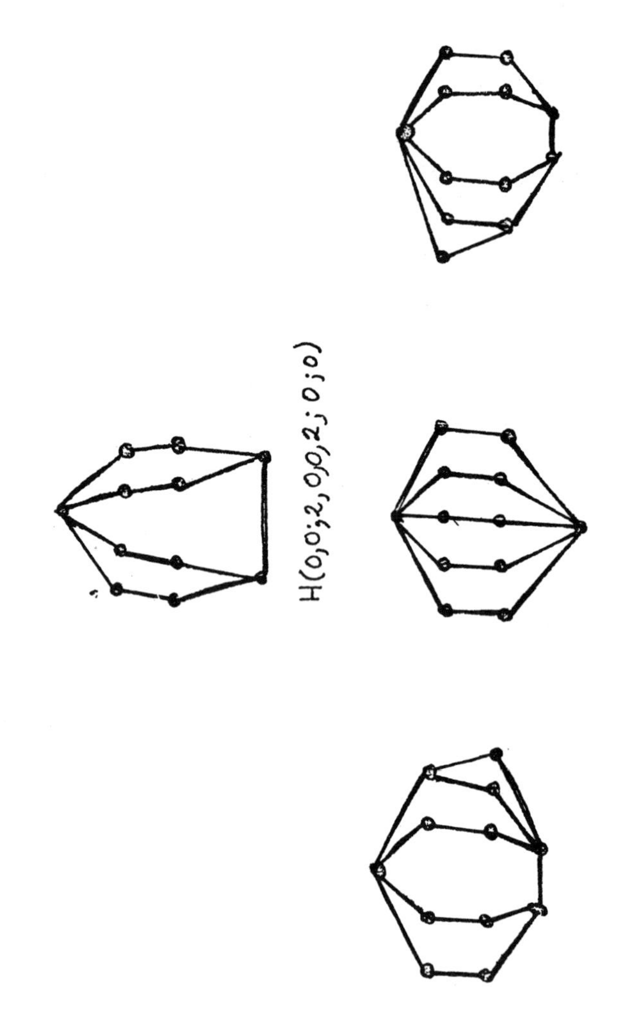

Figure 6. The minimum minimal blocks of diameter 3 on 11 and 12 points

REFERENCES

1. J. A. Barnes, Social Networks. Addison-Wesley Module 22, Reading (1973).
2. B. Bollobás, Extremal Graph Theory. Academic, London (1978).
3. B. Bollobás and F. Harary, Extremal graphs with given diameter and connectivity, Ars Combinatoria, 1 (1976) 281-296.
4. G. A. Dirac, Minimally 2-connected graphs. J. Reine Angew. Math. 228 (1967) 204-216.
5. F. Harary, Graph Theory. Addison-Wesley, Reading (1969).
6. U.S.R. Murty, Extremal nonseparable graphs of diameter 2. Proof Techniques in Graph Theory (F. Harary, ed.) Academic Press, New York (1969) 111-118.
7. M. D. Plummer, On minimal blocks. Trans. Amer. Math. Soc. 134 (1968) 85-94.

A FORBIDDEN SUBGRAPH CHARACTERIZATION OF INFINITE GRAPHS OF FINITE GENUS

Joan P. Hutchinson and Stanley Wagon

Department of Mathematics
Smith College
Northampton, MA 01063

ABSTRACT. We present a characterization of infinite graphs of finite genus. The main result, which generalizes earlier work of Moore and Wagner for the plane, can be phrased as a forbidden subgraph characterization of those graphs which embed on the sphere plus k handles.

1. INTRODUCTION

It is well known that every finite graph can be embedded on a sphere with k handles, for some finite nonnegative integer k, but the same assertion for infinite graphs has not been considered, let alone settled, except for the plane, $k = 0$. This lack of study may be due to the difficulty of obtaining a Kuratowski-type theorem for nonplanar surfaces; but by looking beyond the embedding problem for finite graphs, we shall see that the case of infinite graphs is quite tractable. There are some obvious conditions that an infinite graph must satisfy

---

(Written while visiting the University of Colorado, Boulder, Colorado).

if it is to embed on a surface of finite genus, and we shall show that these conditions, together with the non-existence of a subgraph lying in a certain family, are necessary and sufficient for embeddability. Moreover, the forbidden family is the same for all surfaces.

We use the notation and terminology of [15]. By an infinite graph we mean one with an infinite (not necessarily countable) number of vertices; the term "graph" will refer to either a finite or infinite simple graph. We shall say that a graph has <u>finite genus</u> if, for some $k < \infty$, it embeds (using Jordan arcs with no edge crossings) on the sphere with k handles. For such graphs G, the <u>genus</u> of G is the smallest k such that G embeds on the sphere with k handles.

If G has genus k, then it is clear that G must have continuum or fewer vertices and that every subgraph of G must have genus at most k. The following theorem gives a characterization of graphs of genus k. Its proof will be given after three additional results, which serve as lemmas.

Theorem 1. A graph G has genus k if and only if:

(1) G has at most continuum many vertices,

(2) k is the maximum genus of all finite subgraphs of G, and

(3) G has at most countably many vertices of degree at least three.

We shall see (Corollary 5) that these conditions are equivalent to a forbidden subgraph criterion. Note that condition (2) requires knowledge of the genus of the finite subgraphs of G. Of course, it is usually difficult to determine whether a finite graph embeds on a given surface. Kuratowski's theorem [5] may be used for the plane and recent work [1,3] for the projective plane, but no analog of these results is yet known for other surfaces.

The planar (genus 0) version of Theorem 1 has been known for some time and is due to Moore [6,7] and Wagner [14]. Precisely, Moore proved in 1928 that no planar graph can contain uncountably many vertices of degree three or more, and Wagner, unaware of Moore's work, in 1967 proved Theorem 1 with $k = 0$ and condition (2) replaced by: (2') G contains no homeomorph of K_5 or $K_{3,3}$. We give a new, simpler proof of Moore's result and then deduce that no graph of finite genus can contain uncountably many vertices of degree three or more. Then we prove the sufficiency of conditions (1) - (3) for all k, following the ideas of Wagner's proof. The planar case has also been studied and extended in other ways [4,11] which we shall discuss in Section 3.

We would like to thank our colleague J. Henle for stimulating discussions on these topics. We would also like to thank R.H. Bing, V. Klee, J. Mycielski and G.S. Young for helping us trace the history relevant to this work.

2. THE MAIN RESULTS

We begin with results which show that no graph of finite genus can have too many vertices of "large" degree. The proofs of Theorem 2 in [6] and [14] are indirect and use limit point arguments, which are avoided in the direct approach below.

Theorem 2. (Moore [6,7]) A planar graph cannot have uncountably many vertices of degree three or more.

Proof. Suppose G is a graph, embedded in the plane, with vertex set V. Let $B = \{v \in V : \text{degree}(v) \geq 3\}$, and let D be the countable dense subset of the plane consisting of points both of whose coordinates are rational. For any $v \in B$ one can find $Q \in D$ and $n \in \underline{N}$ such that the circle of radius $1/n$ centered at Q satisfies:

(1) v is in the interior of C,

(2) There are three edges (Jordan arcs) J_1, J_2, J_3 emanating from v which intersect C.

Let P_i be the point on C where J_i (of (2) above) first crosses C; the continuity of J_i guarantees the existence of P_i. The three Jordan arcs, truncated at P_i, divide the interior of C into three regions; choose points Q_1, Q_2, Q_3 in D so that one Q_i lies in each of these three regions. It is easy to see that distinct vertices in B yield distinct 5-tuples (n, Q, Q_1, Q_2, Q_3). Since $N \times D^4$ is countable, so is B. □

One perhaps surprising consequence of this result is that there are no uncountable three-connected planar graphs. Thomassen [12, Lemma 7.3] has proved that there are no 2-connected uncountable planar graphs with a 2-basis.

Theorem 3. A graph of genus k cannot have uncountably many vertices of degree three or more.

This result is an easy consequence of the fact that the sphere plus k handles can be covered by a finite number of open sets, each homeomorphic to a disc. It follows from Theorem 2 that each such open set can contain at most countably many vertices of degree three or more. For the same reason, this result holds for graphs embedded on nonorientable surfaces of finite Euler characteristic.

Theorems 2 and 3 lead to the family $\mathcal{T}$ of forbidden subgraphs. Precisely, $\mathcal{T}$ is the class of all bipartitie graphs in which one part of the bipartition consists of $\aleph_1$ vertices, each of degree three. For example, the vertex-disjoint union of $\aleph_1$ copies of $K_{1,3}$ is in $\mathcal{T}$ as is $K(\aleph_1, 3)$. In fact, every graph in $\mathcal{T}$ contains uncountably many edge-disjoint $K_{1,3}$'s.

Proposition 4. G has uncountably many vertices of degree at least three if and only if some subgraph of G lies in $\mathcal{T}$.

Proof. For any graph G, let B(G) consist of those

vertices of G whose degree is at least three. One direction of the proposition is clear, so assume G is a graph such that $B(G)$ is uncountable. We claim that for some $v_0 \in B(G)$ and G_1 the induced subgraph on $V - v_0$, $|B(G) \setminus B(G_1)| \leq 4$. In other words, the deletion of v_0 turns at most three vertices of $B(G)$ into vertices of degree two. To find v_0, choose any $v \in B(G)$. If v fails to be as claimed, then v has at least four neighbors in $B(G)$, each of which has degree exactly three. Any of these neighbors may be selected as v_0 since its removal reduces the degree of only its neighbors.

Now, to construct the desired bipartite subgraph T of G, let v_0 be as claimed above and put it into one set A of the bipartition of T. Select three neighbors of v_0 and put them in B, the other set of the bipartition, and let T contain the edges from v_0 to these three vertices. Then repeat the procedure, starting with G_1. Since $B(G_1)$ is uncountable, we obtain v_1 as we did v_0, and add v_1 to A, add three of its neighbors (not necessarily distinct from the vertices already in B) to B and add three connecting edges to T. We may use transfinite induction to continue the construction of T so that A consists of $\{v_\alpha : \alpha < \omega_1\}$. The key point is that at the α'th stage, $B(G_\alpha)$ is uncountable where G_α is the induced subgraph of G with vertex set $G \setminus \{v_\beta : \beta < \alpha\}$; this follows from the fact that the removal of each v_β eliminates at most four vertices from $B(G_\beta)$. □

We now complete the proof of Theorem 1; the proof of sufficiency adapts many of Wagner's ideas for the planar case.

Proof of Theorem 1. The necessity of the conditions is clear from previous discussion and Theorems 2 and 3. If G satisfies conditions (1) - (3), we shall see that it embeds on S_k, the sphere plus k handles. We partition (the vertices of) G into two (induced) graphs H_1 and H_2 where H_1 is the union of all components of G with maximum degree two and H_2 is the union of all other components. Thus H_1 is the union of isolated vertices, cycles and (finite or infinite) paths. These will be embedded, at the end, in some small patch of the surface. Delete from H_2 all vertices of degree one or two and their incident edges, but join two remaining vertices by an edge if they are not now adjacent, but in H_2 were joined by a path of vertices of degree two. Let H_2' denote the graph resulting from this alteration of H_2.

By assumption, there are at most countably many vertices in H_2', since all these vertices originally had degree at least three. Note that the transformation from H_2 to H_2' preserves the property that every finite subgraph embeds on S_k. For any graph F, let DF denote the "double" of F, i.e., the multigraph obtained by replacing each edge of F by a pair of edges. By a di-gon embedding of DF we mean an embedding of DF on S_k with the property that if e and e' are the two edges on S_k connecting a pair of vertices, then e and e' bound a contractible di-gon (2-cell) which is disjoint from all vertices and edges of the embedding. If F is finite and embeds on S_k then there is a di-gon embedding of DF on S_k: consider the finitely many edges and, one at a time, replace them by two disjoint edges chosen close enough to each other to avoid all other parts of the graph.

Our overall plan is to obtain a di-gon embedding of DH_2', transform it to an embedding of H_2, and then expand the latter to the desired embedding of G. The existence of a di-gon embedding of DH_2' follows from the following observation.

Lemma. If H is any countable graph with the property that any finite subgraph of H embeds on S_k, then there is a di-gon embedding of DH on S_k.

Proof. It is well-known that if every finite subgraph of a countable graph embeds on S_k then the graph embeds there too. See [2] where this is proved via the König Infinity Lemma or [8,9] for an interesting approach based on the Tychonoff Product Theorem. Both of these methods work for multigraphs, and it is easy to see that if only di-gon embeddings of the finite subgraphs of DH are considered (such exist by the remarks above), then the ultimate embedding is a di-gon embedding as well. □

We apply the lemma to the graph H_2' and transform the resulting di-gon embedding of DH_2' into an embedding of H_2 as follows. For each adjacent pair of vertices $u, v \in DH_2'$ choose a homotopy $\{h_t\}$ of the two edges joining u and v on S_k such that the family of paths $\{h_t\}$ is pairwise interior-disjoint and each h_t lies in the di-gon bounded by the two edges. Now, $\{h_t : 0 < t < \frac{1}{3}\}$ can be used to embed the paths from u to v in H_2 that consist of vertices of degree 2, and $\{h_t : \frac{1}{3} < t < \frac{2}{3}\}$ can be used to embed any such paths in H_2 from u (or v) to a vertex of degree one. Finally, pick one of the homotopies above and embed H_1 in the interior of the region bounded by $h_{2/3}$ and h_1. Erasing all multiple edges and any edges of H_2' that were not in H_2 turns this into the desired embedding of G.

Note that the same proof holds for G a multigraph with no more than $\underline{c}$ (continuum) edges. We also observe that the result holds for graphs on nonorientable surfaces of finite Euler characteristic since the methods of [2] and [8] hold for nonorientable surfaces as well.

To carry our forbidden subgraph characterization one step further, let $\mathcal{G}$ be the class of all graphs that consist of just $\underline{c}^+$ isolated vertices and let $\mathcal{S}_n$ $(n \geq 0)$ consist of all finite graphs of genus n.

Corollary 5. A graph G embeds on the sphere with k handles $(k \geq 0)$ if and only if G contains no subgraph in $\mathcal{G} \cup \mathcal{T} \cup \bigcup_{n>k} \mathcal{S}_n$.

3. EXTENSIONS AND QUESTIONS

A graph G has infinite genus if G embeds on no S_k, $k < \infty$. (This definition differs slightly from that in [15, p. 71].) Thus all members of $\mathcal{T}$ have infinite genus as does $K_{\aleph_0}$ or any graph with more than $\underline{c}$ vertices. Since every graph with at most $\underline{c}$ vertices is a subgraph of the complete graph with $\underline{c}$ vertices, it is worth noting that the latter does embed in $\underline{R}^3$ and furthermore with every edge a straight line segment, by placing vertices on the points (t,t^2,t^3), $t \geq 0$; see [15, p. 51].

A planar graph is said to have a straight line representation if it is isomorphic to a planar graph with each edge a straight line segment; Thomassen [11] has shown that every planar graph has such a representation. By analogy, we say that a graph of genus k has a straight line representation on the sphere with k handles if it is isomorphic to a graph drawn in a 4k-gon (with appropriate pairs of sides identified) with each edge in the interior of the 4k-gon a straight line segment and each edge crossing the boundary a finite sequence of straight line segments, each joining an end vertex to the boundary of the polygon or joining two points on the boundary. In fact, a result of Thomassen [12, Theorem 5.1] shows that every finite, 3-connected graph of genus k has not only a straight line representation, but even a convex one inside a 4k-gon, that is, a representation in which each face is a convex polygon. We conjecture that every graph of genus k has a straight line representation on the sphere plus k handles.

Considerable study has also been made of the conditions under which a planar graph has an embedding which is free of vertex or edge accumulation points (VAP-free or EAP-free respectively). Halin [4] has characterized the locally finite planar graphs having a VAP-free representation and Schmidt [10] has extended this to arbitrary infinite graphs. A variety of related results can be found in [13]. But the sphere with k handles is compact, so one obtains quite simply that a graph of genus k ($k \geq 0$) has a VAP-free (respectively, EAP-free) embedding on the sphere plus k handles if and only if G has a finite number of vertices (edges).

Finally, we mention a topological generalization of Moore's result to higher dimensions. By a _standard umbrella_ in $\underline{R}^n$ ($n \geq 3$) we mean $\{(x_1,\ldots,x_{n-1},0) : \Sigma x_i^2 \leq 1\} \cup \{(0,\ldots,0,x_n) : -1 \leq x_n \leq 0\}$; the terminology arises from the shape of this set in $\underline{R}^3$. By an _umbrella_ in $\underline{R}^n$ we mean any subset of $\underline{R}^n$ that is homeomorphic to the standard umbrella. Then Moore's result can be generalized to show that any family of pairwise disjoint umbrellas in $\underline{R}^n$ ($n \geq 3$) is countable. This was proved by Young [16], although Zarankiewicz [17] had earlier proved the result for umbrellas that are similar to the standard umbrella.

REFERENCES

1. D.S. Archdeacon, A Kuratowski theorem for the projective plane. J. Graph Theory 5(1981) 243-246.
2. G.A. Dirac and S. Schuster, A theorem of Kuratowski. Indag. Math. 16(1954) 343-348.
3. H.H. Glover and J.P. Huneke, The set of irreducible graphs for the projective plane is finite. Discrete Math. 22(1978) 243-256.
4. R. Halin, Zur häufungspunktfreien Darstellung abzählbarer Graphen in der Ebene. Arch. Math. 17(1966) 239-243.
5. K. Kuratowski, Sur le probleme des courbes gauches en topologie. Fund. Math. 15(1930) 271-283.
6. R.L. Moore, Concerning triods in the plane and the junction points of plane continua. Proc. Nat. Acad. Sci., U.S.A. 14(1928) 85-88.
7. R.L. Moore, Concerning triodic continua in the plane. Fund. Math. 13(1929) 261-263.
8. J. Mycielski, Some remarks and problems on the colouring of infinite graphs and the theorem of Kuratowski. Acta Math. Acad. Sci. Hungar. 12(1961) 125-129.
9. J. Mycielski, Correction to my paper on the colouring of infinite graphs and the theorem of Kuratowski. Acta Math. Acad. Sci. Hungar. 18(1967) 339-340.

10. R. Schmidt, Doctoral dissertation. Universität Hamburg, 1976.
11. C. Thomassen, Straight line representations of infinite planar graphs. J. London Math. Soc. 16(1977) 411-423.
12. C. Thomassen, Planarity and duality of finite and infinite graphs. J. Combin. Theory. B 29 (1980) 244-271.
13. C. Thomassen, Kuratowski's theorem. J. Graph Theory 5(1981) 225-241.
14. K. Wagner, Fastplättbare Graphen. J. Combin. Theory 3(1967) 326-365.
15. A.T. White, Graphs, Groups and Surfaces, North-Holland, Amsterdam (1973).
16. G.S. Young, A generalization of Moore's theorem on simple triods. Bull. Amer. Math. Soc. 50(1944) 714.
17. K. Zarankiewicz, Über doppeltzerlegende Punkte. Fund. Math. 23(1934) 166-171.

EMPIRE MAPS

Brad Jackson and Gerhard Ringel

Department of Mathematics
University of California
Santa Cruz, CA 95064

ABSTRACT. Heawood generalized his Map Color Problem to empire maps. Solutions are given here for the projective plane and some other surfaces.

The Four Color Theorem says that the vertices of each planar graph can be colored by four colors, or that the countries of each map on the plane are colorable by four colors. Heawood in 1890 could not prove this theorem so he generalized the concept. He did it in two directions. First -- take instead of the plane any other surface, for instance the torus.

Second -- consider maps where certain collections of nonadjacent countries are called empires and each country belongs to exactly one empire. These maps will be called empire maps. Of course in a coloring of an empire map each component of an empire should have the same color and any two adjacent countries have to have different colors.

In the 90 years after Heawood's paper was published, mathematicians worked on the first aspect until the question was finally solved completely. But they almost entirely neglected the second aspect.

Let S be a surface and M a positive integer. By $\chi(S,M)$ we denote the minimum number of colors necessary and sufficient to color every empire map on S, where each empire had no more than M components. This number is called the M-pire chromatic number of S. We call an empire with exactly m components an m-pire.

Before we report general results we consider the special case where S is the plane or the projective plane. The following two theorems [10, p. 24] are often very helpful in coloring problems.

Theorem 1. If G is a planar or projective planar graph, and d(G) is the average of the degrees of the vertices in G then $d(G) < 6$.

Theorem 2. Let T be a set of graphs and assume that for every graph $G \in T$ each subgraph of G is also in T and that $d(G) < h$ for each $G \in T$. Then the vertices of G are colorable by h colors for each $G \in T$.

If we apply Theorems 1 and 2 to planar graphs, we obtain the result that every planar graph is colorable by 6 colors. This is of course very disappointing. However for other problems Theorem 2 will be very sharp.

Given an empire map we can define a graph where each country, respectively each empire, is represented by a vertex and two vertices are adjacent if and only if the two represented countries, respectively empires, are adjacent in the map. This graph is called the country graph, respectively the empire graph, of the empire map. It is clear that one obtains the empire graph from the country graph by identifying those vertices which represent the same empire. Coloring an empire map is then equivalent to coloring its empire graph.

We consider all empire maps on the plane or projective plane where each empire has no more than M components. Then because of Theorem 1 in the corresponding empire graphs the average degree of the vertices is $< 6M$. Then Theorem 2 gives:

$$\chi(\text{plane}, M) \leq 6M, \tag{1}$$

$$\chi(\text{projective plane}, M) \leq 6M. \tag{2}$$

Heawood exhibited a map on the plane with 12 mutually adjacent 2-pires showing that equality holds in (1). Taylor [4] constructed two more maps showing that (1) holds as an equality also for $M = 3$ and 4. For $M \geq 5$ the matter is unsolved. The maps of Heawood and Taylor are very irregular and have no symmetry, and no pattern. There is not much hope of generalizing them. It seems easier to find empire maps on other surfaces. We give here a short sketch of a proof for equality in (2) for each M.

Consider the example $m = 3$ given in Figure 1. It shows an empire map on the projective plane consisting of 18 mutually adjacent empires; 17 of them are 3-pires and one is a 1-pire. One can see that the permutation $(0\ 1\ 2\ \cdots\ 16)$ which replaces i by $i+1$ (mod 17) describes an automorphism of the map. Notice that the labels of the neighbors of the 3-pire 0 are given in three cycles

(5,2,8,x,9,11,1), (3,7,6,15), (4,1,16,10,13,12,14)

or written in the form

(5,2,8,x,-8,-6,1), (3,7,6,-2) (4,1,-1,-7,-4,-5,-3)

These three cycles we obtained by travelling

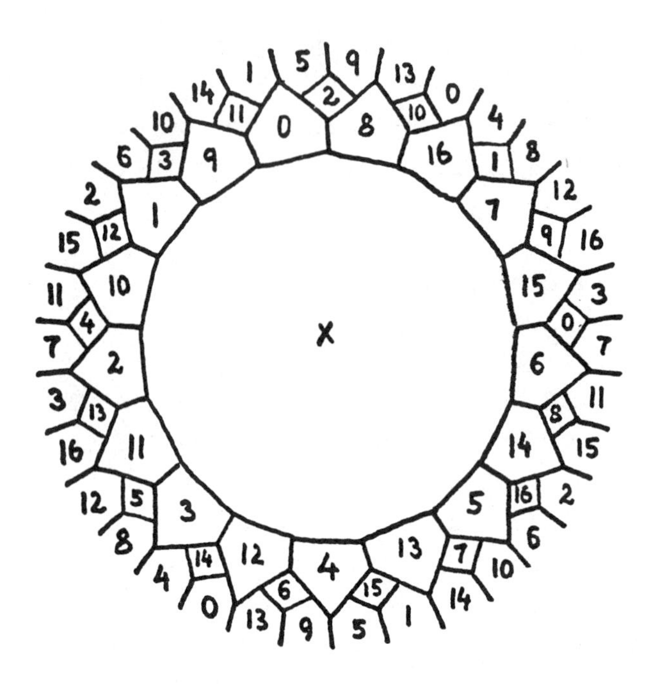

Figure 1.

through the "cascade" of Figure 2. A cascade is a modification of a current graph -- a construction device. For details see Ringel [11, p. 13]. Notice that at each vertex of degree 3 in Figure 2 Kirchoff's Current Law is true: The sum of the incoming currents equals the sum of the outgoing currents. This law guarantees that the constructed map will have cubic vertices only.

The edge with current 4 is a "broken" edge; this makes the current graph a cascade and the generated map non-orientable. Using Euler's formula one can prove that the map defines the projective plane.

The next example m = 7 in Figure 3 shows how the concept can be generalized. For even integers m the cascade is slightly different. These were the ideas used to prove that equality holds in (2).

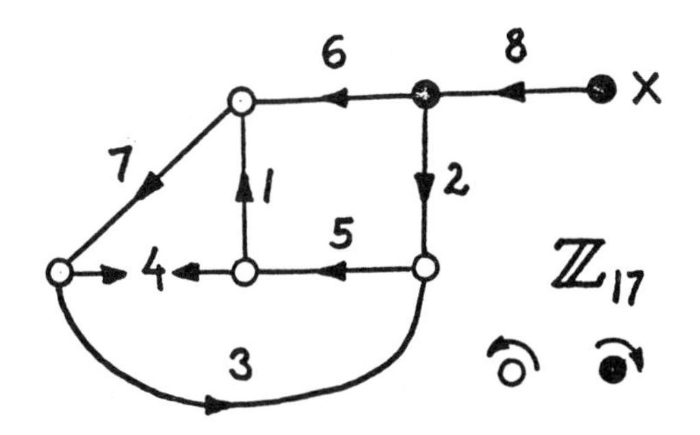

Figure 2.

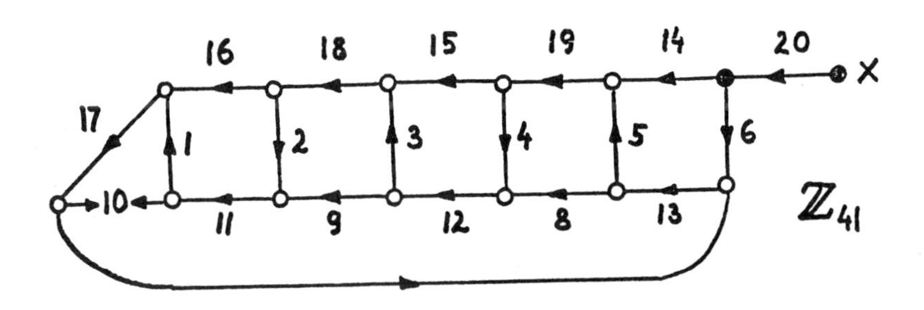

Figure 3.

Now we report the present situation of the general problem of determining the number of colors $\chi(X,M)$, which is sufficient and necessary to color every empire map on the surface S where each empire has no more than M components. Let E be the Euler characteristic of the surface S.

Heawood in 1890 showed the inequality

$$\chi(S,M) \leq \left\lfloor \frac{1}{2}\left(6M + 1 + \sqrt{(6M+1)^2 - 24E}\right)\right\rfloor \tag{3}$$

for every surface S and every natural number M with one exception (M = 1, S = sphere). This exception is the Four Color Theorem which was proved by Haken and Appel [1,2]. So (3) holds in general. Heawood conjectured that equality always holds in (3).

In the following cases equality in (3) has been proven.

(a) M = 1, S is a non-orientable surface ≠ Klein's bottle. Ringel [9], 1954. If S is the Klein bottle and M = 1, then (3) does not hold as an equality. Franklin [3], 1934.

(b) M = 1, S is orientable. Ringel, Youngs and others [11], 1968.

(c) M = 2, S is the sphere. Heawood [5], 1890.

(d) M = 3 or 4, S is the sphere. Taylor [4], 1890.

(e) S = torus. Taylor [12], 1982.

(f) S is the projective plane. Jackson, Ringel [6], 1983.

(g) S is non-orientable and the right-hand side of (3) is congruent to 1, 4, or 7 (mod 12). Jackson, Ringel [8], 1983.

(h) S is orientable, M is even, and the right-hand side of (3) is congruent to 1 (mod 12). Jackson, Ringel [8], 1983.

(i) S is orientable, M is odd, and the right-hand side of (3) is congruent to 4 or 7 (mod 12). Jackson, Ringel [8], 1983.

REFERENCES

1. K. Appel and W. Haken. The existence of unavoidable sets of geographically good configurations, Illinois J. Math. 20 (1976) 218-247.

2. K. Appel, W. Haken and J. Koch. Every planar map is four-colorable, Illinois J. Math. 21 (1977) 429-567.

3. P. Franklin. A six color problem, J. Math. Phys. 13 (1934), 363-369.

4. M. Gardner. Mathematical Games, Scientific American. Feb. 1980.

5. P. J. Heawood. Map Colour Theorem, Quart. J. Math. 24, (1890) 332-333.

6. B. Jackson and G. Ringel. Maps of m-pires on the Projective Plane, to appear in Discrete Math.

7. B. Jackson and G. Ringel. The splitting number of the complete graph, in preparation.

8. B. Jackson and G. Ringel. Heawood's Empire Problem, (to appear) J. Combin. Theory.

9. G. Ringel. Bestimmung der Maximalzahl der Nachbargebiete auf nichtorientierbaren Flächen. Math. Ann. 127 (1954), 181-214.

10. G. Ringel. Färbunsprobleme auf Flächen und Graphen Berlin: VEB Deutscher Verlag der Wissenschaften 1959.

11. G. Ringel. Map Color Theorem. Band 209, Springer-Verlag, Berlin 1974.

12. H. Taylor. The m-pire chromatic number of the torus is 6m + 1, J. Graph Theory, to appear.

SPLITTINGS OF GRAPHS ON SURFACES

Brad Jackson and Gerhard Ringel
University of California
Santa Cruz, CA 95064

ABSTRACT. Let $\sigma(G,S)$ be the smallest number of vertex splittings needed to transform the graph G into a graph embeddable on S. A lower bound for $\sigma(G,S)$ is derived. The lower bound is shown to be achieved for all complete bipartite graphs on any surface S. If S is the plane, equality is shown to hold for complete graphs K_n when $n \equiv 2 \pmod 3$. It is conjectured that equality always holds on any surface except when n = 6,7,9.

Given a graph H we construct a new graph G with one less vertex by replacing two vertices u and v of H by a single vertex which is adjacent with all the vertices of H that are adjacent with either u or v (no loops or multiple edges are formed). We say that G is obtained from H by a *vertex identification*. The *splitting number* $\sigma(G,S)$ of G on a surface S is the smallest number k so that G can be obtained from some graph embeddable on S by k vertex identifications. We denote by $\sigma(G)$ the planar splitting number of G. The reverse of the process of vertex identification is called *vertex splitting.* Thus the *splitting* number $\sigma(G,S)$ is the smallest number of vertex splittings needed to transform G into a graph embeddalbe on S. Of course $\sigma(G,S)$ is zero if and only if G is embeddable on S. So the splitting number of G on S is a measure to how close a graph G is to being embeddable on S.

We say that H is <u>a splitting</u> of G on S if H is embeddable on S and if H can be obtained from G by vertex splittings. In Figure 1 we have a planar splitting of

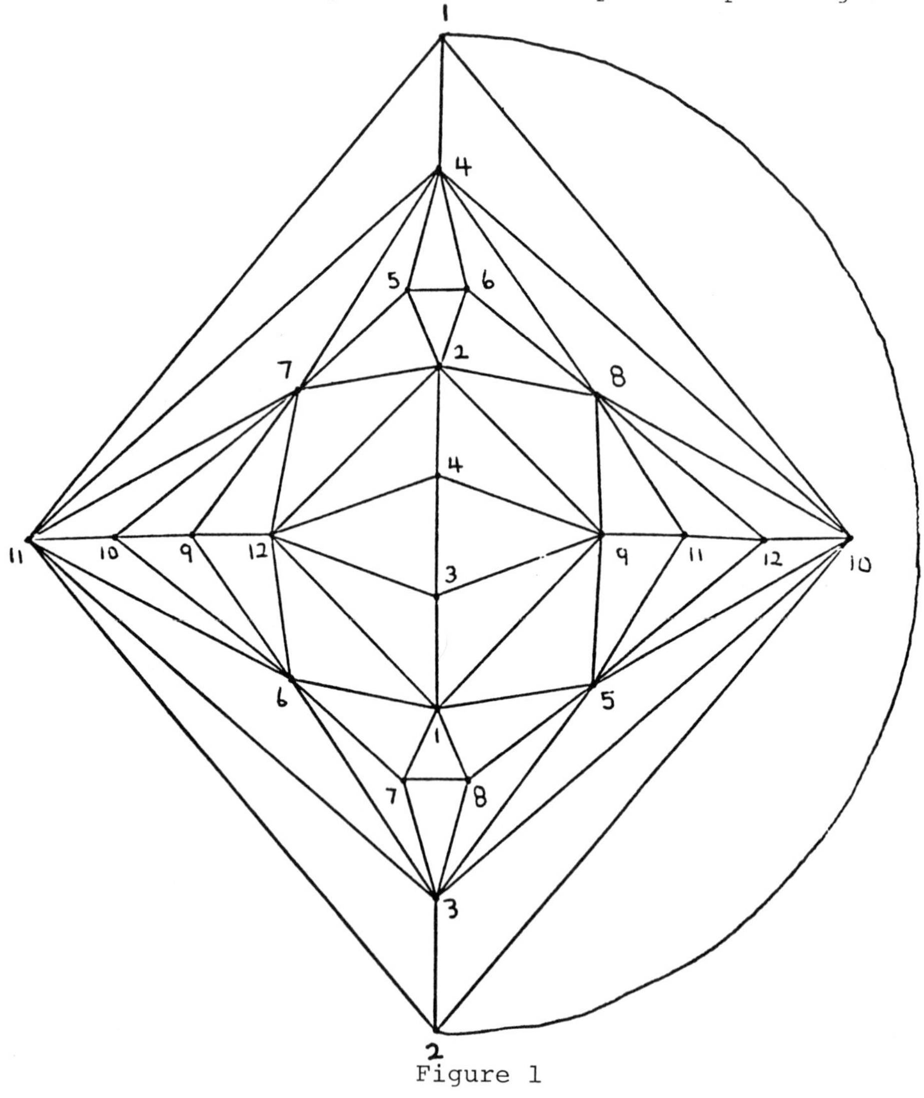

Figure 1

K_{12}. Since this graph can be transformed to K_{12} by 12 vertex identifications, which consist of identifying pairs of like numbered vertices, we know that $\sigma(K_{12}) \leq 12$. We will also derive a lower bound for splitting numbers, mainly using Euler's formula, which will enable us to show that in fact $\sigma(K_{12}) = 12$. Dual to this particular splitting of K_{12} we have a planar map of 12 mutually adjacent "2-pires", [1]. Conversely some important "symmetrical splittings" of K_n will be obtained from maps of n mutually adjacent "empires" which we have constructed in [1]. For the definition of empires and empire maps see [5].

First we compute a lower bound for $\sigma(G,S)$. Let G be an arbitrary graph with p vertices, q edges, and with girth g. If $\sigma(G,S) = s$ then there exists a splitting H which can be obtained from G by s vertex splittings. Since H has p+s vertices, $q^* > q$ edges, and the girth of H is at least g (possibly after some duplicate edges are deleted), a well-known argument shows that

(0) $\quad 2q^*/g - q^* + (p+s) \geq E(S)$

(1) $\quad s \geq ((g-2)q^* - gp)/g + E(S)$

(2) $\quad s \geq ((g-2)q - gp)/g + E(S)$

(3) $\quad s \geq \lceil((g-2)q - gp)/g + E(S)\rceil$

Also it is known that equality holds in (1) if and only if every face of the embedding of H in S is a polygon with g sides. Thus equality holds in (2) if and only if H is a splitting of G on S, with the same number of edges as G, which can be embedded on S with all faces having g sides. We call such a splitting H, a <u>perfect splitting</u> of G on S. When the right side of (2) isn't an integer but H is a splitting of G on S for which equality holds in (3) then we call H an <u>almost perfect splitting</u> of G on S. We call a splitting of G which has a triangular embedding (respectively quadrilateral embedding) on S a triangular splitting (respectively quadrilateral splitting) of G on S.

Consider first the complete bipartite graph $K_{m,n}$ with m red vertices and n blue vertices, together with all mn edges joining vertices of different colors. For splitting numbers of $K_{m,n}$ we proved the following theorem in [2].

Theorem A. $\sigma(K_{m,n}, S) = \{\max 0, E(S) - 2 + (m-2)(n-2)/2\}$

To prove Theorem A we first constructed a perfect or almost perfect splitting of $K_{m,n}$ on the plane. For example, ignoring the dotted lines, Figure 2 shows a

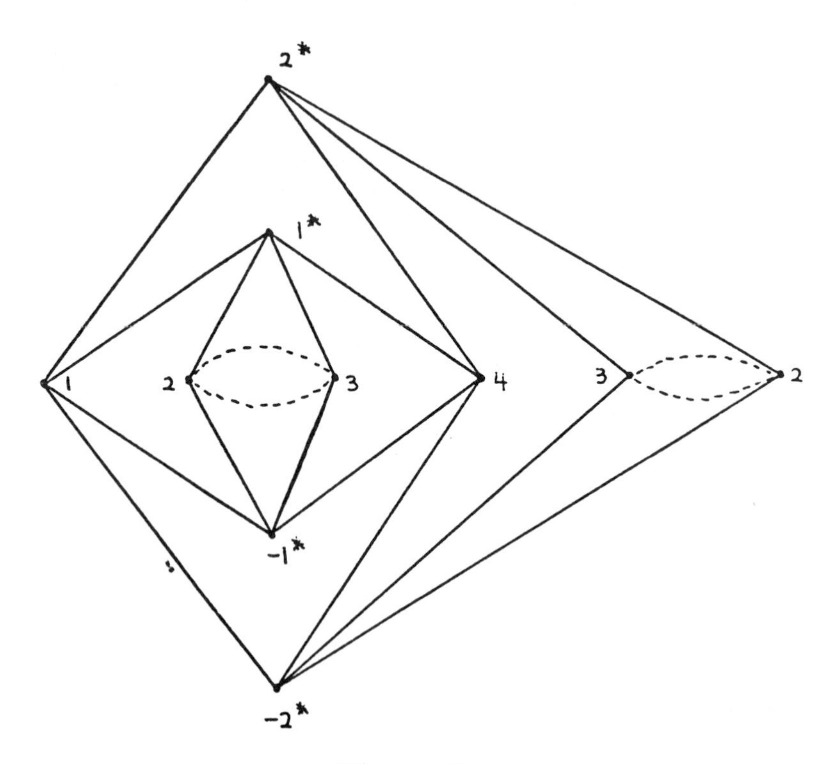

Figure 2

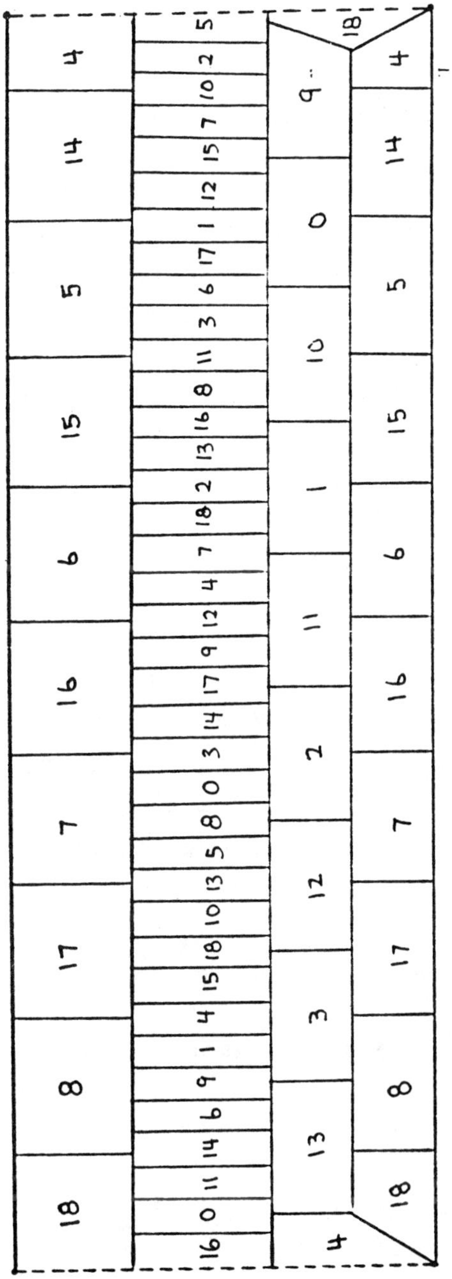

18 8 17 7 16 6 15 5 14 4

16 0 11 14 6 9 1 4 15 18 10 13 5 8 0 3 14 17 9 12 4 7 18 2 13 16 8 11 3 6 17 1 12 15 7 10 2 5

4 13 3 12 2 11 1 10 0 9 18

18 8 17 7 16 6 15 5 14 4

Figure 3

perfect splitting of $K_{4,4}$ on the plane. We then constructed quadrilateral splittings of $K_{m,n}$ on other surfaces by adding handles, respectively crosscaps, which identified two, respectively one, pair of vertices. For example one can add pairs of nonintersecting arcs between vertices 2 and 3 in quadrilaterals [1* .3,-1*,2] and [2*,3,-2*,2] as in Figure 2. Then excise the interior of the two lunes formed and identify the two boundaries in one of two essentially different ways so that the two vertices numbered 2 and also the two vertices numbered 3 are identified. In this way we add a handle to the surface and since two pairs of vertices are identified, the resulting embedding still gives a perfect or almost perfect splitting of $K_{m,n}$.

In a similar way one can add a crosscap to the surface, by first cutting along an edge from 3 to 4 in quadrilateral [1*,3,-1*,4] continuing from 4 to 3 in quadrilateral [2*,4,-2*,3]. By reidentifying along this cut after a 180° twist, one identifies the pair of vertices numbered 3 and at the same time adding a crosscap to the surface. Using these and similar operations one can construct perfect or almost perfect splittings of $K_{m,n}$ on any surface up to and including orientable and nonorientable genus embeddings of $K_{m,n}$ (no vertices split).

We have attempted to compute the splitting numbers of complete graphs in a similar manner, so far with less complete results. First we prove the following theorem about planar splittings of K_n.

Theorem 1. $\sigma(K_n) = \lceil (n-3)(n-4)/6 \rceil + e(n)$ where

a) $e(n) = 0$ if $n \equiv 2 \pmod 3$

b) $e(n) = 0$ or 1 if $n \equiv \pmod 3$

Conjecture 1. For $n \geq 10$, $\sigma(K_n) = \lceil (n-3)(n-4)/6 \rceil$.

We have shown that Conjecture 1 is true for $10 \leq n \leq 36$ and in [3] we show that it is true for $n \geq 80$. It is interesting to note that for $n = 6,7,9$ no perfect splitting of K_n exists, so $e(n) = 1$ is best possible. For $n = 6$, respectively $n = 7$, the lower bound for $\sigma(K_n)$ on the right side of (3) is 1, respectively 2. Thus in either of these cases a perfect splitting of K_n would have five mutually adjacent vertices which is impossible. For $n = 9$ the proof that no perfect splitting exists is more complicated.

Proof of Theorem 1. Let H be a splitting of G on S with $s+e$ vertex splittings where s is the right side of (3), then e is called the excess of H. Starting from some empire maps on the plane we are able to construct the appropriate splittings of K_n to show that Theorem 1 holds. We do so by exhibiting almost perfect splittings of K_n on the plane when $n \equiv 2 \pmod 3$ and triangular splittings of excess one when $n \equiv 0,1, \pmod 3$. In either case the proof only holds for $n \geq 37$. We have shown that the theorem holds for $n \leq 36$ but we omit the details.

Consider the current graphs in Figures 4 and 5. As we have seen in [1], these current graphs can be used to construct maps of $6m+1$ mutually adjacent m-pires (empires with m components) on the Klein bottle and thus the duals of these maps are perfect splittings of K_{6m+1} on the Klein bottle. For $m = 3$ the map, which we call the 3-pire piano map, is shown in Figure 3. But to prove Theorem 1 we need to construct splittings of K_n on the plane. So first we make some minor alterations of these maps to obtain some planar empire maps.

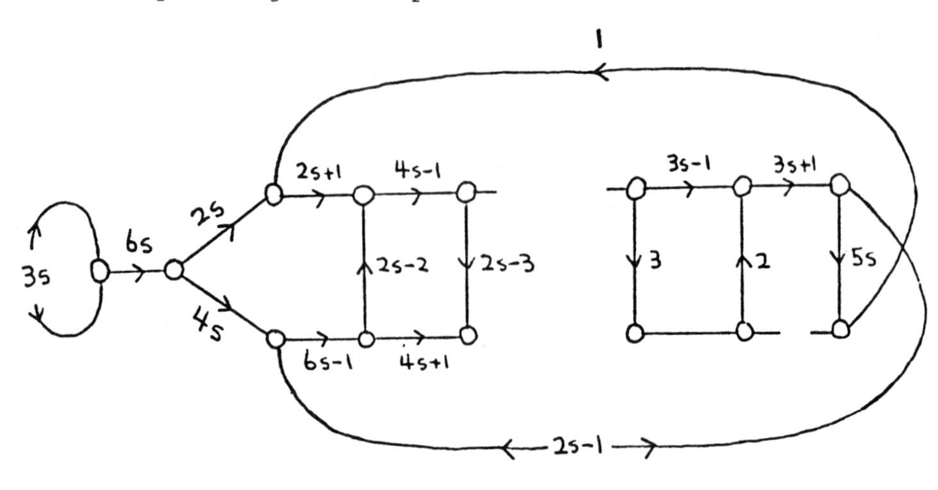

Figure 4

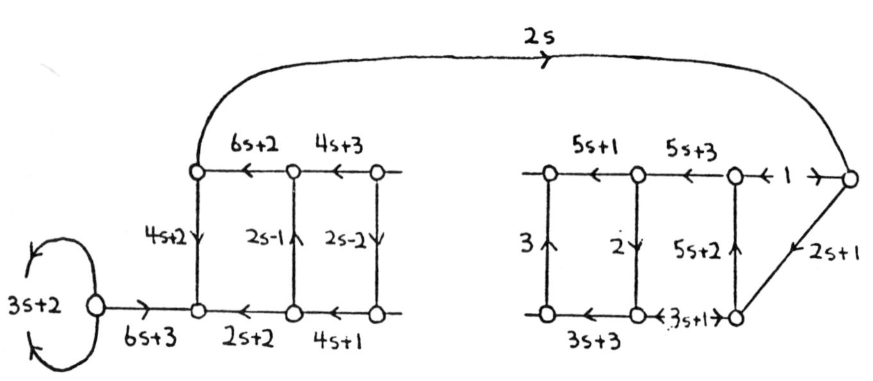

Figure 5

First look at the current graph in Figure 4 with currents in Z_{12s+1}. This current graph has $m = 2s$ circuits. Of these circuits, look in particular at the three circuits $(-6s,-3s,3s,6s,4s,-(2s-1),3s+1,2,-(5s-1),$ $1,-2s)$, $(5s,5s-1,5s+1, \ldots, 4s+1,6s-1,2s-1)$, and $(-5s, -(3s+1), -(3s-1), \ldots, -(4s-1),-(2s+1),-1)$. Each empire z in Z_{12s+1} has countries z_1, z_2, and z_3, one corresponding to each of the three circuits. In circuit

1 we have ... , $-(2s-1)$, $3s+1$, ... , in circuit 2 ... , $2s-1, 5s$, ... , and in circuit 3 ... , $5s$, $-(3s+1)$, ... Thus for every z there will be a vertex where the three countries $(z+2s-1)_1, z_2$, and $(z+5s)_3$ meet. Similarly we can determine that the vertex at the opposite end of the edge between z_2 and $(z+5s)_3$ is $(z+5s-1)_1$. In addition since circuit 1 contains both $3s$ and $-3s$, the countries $(z+5s-1)_1$ and $(z+2s-1)_1$ are adjacent. Let $z = 7s+2$ and cut the map along the edge between $(7s+2)_2$ and 1_3 extending through countries $(9s+1)_1$ and 0_1 to form a cycle which cuts the Klein bottle into a cylinder. Fill in each end of the cylinder with a disk to obtain the map of $12s+1$ empires on the sphere (plane) represented in the upper half of Figure 6. Form a sphere from a map in Figure 6 by identifying the lower half of the boundary. Countries $(9s+1)_1$ and 0_1 from the original map have been divided into two parts and in the new map empire $7s+2$ is no longer adjacent to empire 1. But now we have a map on the plane from which we construct planar maps of $12s+1, 12s+2, 12s+3, 12s+4, 12s+5$, respectively $12s+6$ mutually adjacent empires. From these maps we obtain the appropriate splittings of K_n.

Case 12s+1. We obtain a map of $12s+1$ mutually adjacent empires in the following way. The current graph in Figure 4 has a vertex with rotation $(-1, -2s, 2s+1)$. Thus from the way our original empire map was constructed we see that for every z in Z_{12s+1} there is a vertex where the three empires $z, z-1, z-2s-1$ meet. For $z = 9s+3$ this means that the empires $9s+3, 9s+2, 7s+2$ meet at one point. We denote this point by $[9s+3, 9s+2, 7s+2]$. As in the top half of Figure 7 we add a new country to empire 1, at this point, which is adjacent to each of the three empires 7s+2, 9s+2, and 9s+3. We now have a planar map of 12s+1 mutually adjacent empires with three extra adjacencies since empires 0,9s+1, and 1,9s+2, and 1,9s+3, are each adjacent twice. Thus the dual to this map is a triangular splitting of K_{12s+1} on the plane with excess one.

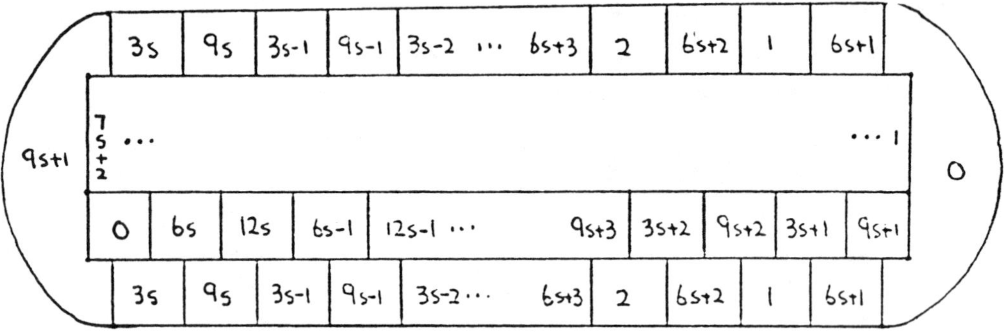

$\mathbb{Z}_{12s+1}$

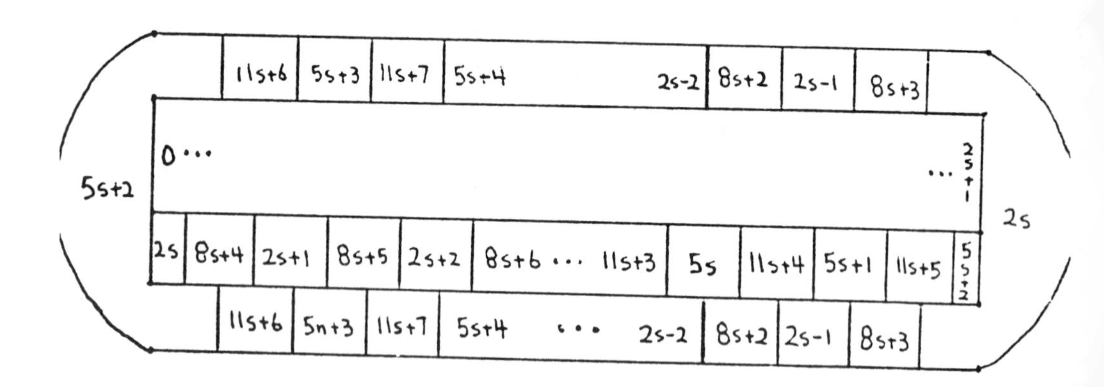

$\mathbb{Z}_{12s+7}$

Figure 6

Case 12s+2. We now add a new empire x_1 to our previous map. As before we add triangular components of x_1 at many vertices of the previous map of 12s+1 mutually adjacent empires. At another vertex of the current graph we have the rotation (-3s,-3s,6s). Thus for every z in Z_{12s+1} we have vertices of our map as pictured in Figure 6 where the empires z,z-3s,z-6s meet. Add triangular components of x_1 at vertices [6s+1,9s+1, 0],[3s,6s,9s],[12s,3s-1,6s-1], ... , and [9s+3,2,3s+2]. Since 3s is relatively prime to 12s+1, all of the multiples of 3s listed above are distinct in Z_{12s+1} and the only remaining elements of Z_{12s+1} are 6s+2,9s+2, 3s+1, and 1. In our previous map we had two adjacencies between empires 1 and 9s+2. At either end of one of these edges between empires 1 and 9s+2 there are the vertices [6s+2,9s+2,1] and [9s+2,1,3s+1]. We denote this edge by [6s+2,9s+2,3s+1,1]. Thus as in the bottom half of Figure 7 we can add a rectangular component of x_1, replacing this edge, which is adjacent to empires 6s+2,9s+2,3s+1, and 1. There is no longer an extra adjacency between 1 and 9s+2 so we have obtained an empire map of 12s+2 mutually adjacent empires with 2 extra adjacencies. The empire map is no longer so symmetrical as it once was since the previous empires had at most 2s+1 components and the new empire x_1 has 4s+1 components. However, most importantly, the dual of this empire map is an almost perfect splitting of K_{12s+2} on the plane.

Case 12s+3. As before we add triangular and rectangular components of a new empire x_2 to our previous map to obtain a map of 12s+3 mutually adjacent empires. Since we have an extra adjacency between empires 0 and 9s+1,

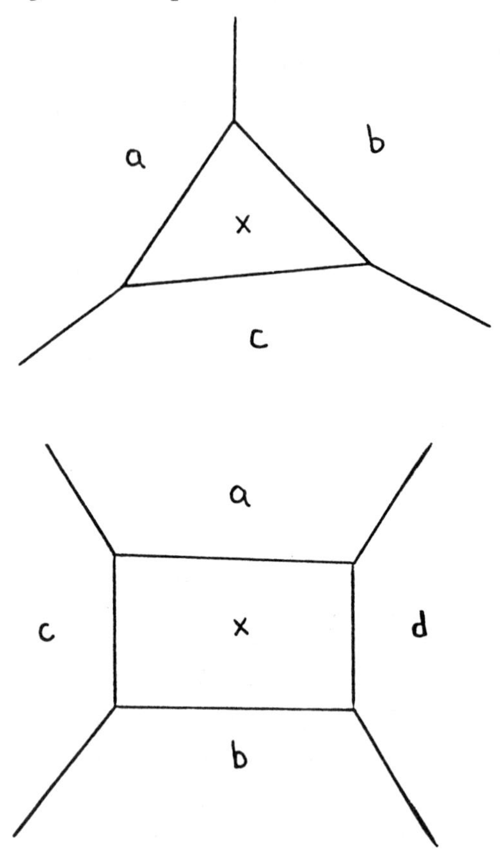

Figure 7

we can add a rectangular component replacing the edge [6s+1,9s+1,0,1]. We also add triangular components of x_2 at [6s,9s,12s],[3s-1,6s-1,9s-1], ...,[2,3s+2,6s+2], [9s+2,3s+1,x_1], and [3s, ,]. We now have a map of 12s+3 mutually adjacent empires on the plane. The two blanks will be extra adjacencies with x_2 and we still have an extra adjacency between 1 and 9s+3 for a total of 3 extra adjacencies. Therefore dual to this map we will have a triangular splitting of K_{12s+3} on the plane with excess one.

Cases 12s+4,12s+5, and 12s+6 are demonstrated similarly adding one new empire each time. To obtain splittings of K_{12s+7}, K_{12s+8}, etc. we start with the empire maps of 12s+7 mutually adjacent (2s+1)-pires on the Klein bottle constructed using the current graph in Figure 5.

Again we make a minor alteration to obtain a map on the sphere that is represented in the lower half of Figure 6. As before we cut along an edge between a country of empire 0 and a country of empire 2s+1 continuing through empires 2s and 5s+2 to form a cylinder. Patching the ends of the cylinder we obtain a map on the sphere which can be obtained from the lower part of Figure 6 by identifying the upper half of the boundary with the lower half. One pair of empires, namely 0 and 2s+1 are no longer adjacent. From this map we construct the maps of 12s+7,12s+8,12s+9,12s+10,12s+11, respectively 12s+12 mutually adjacent empires necessary to obtain the appropriate splittings.

Case 12s+7. Add one component to empire 2s+1 at [0,5s+3, 5s+1] to obtain a planar map of 12s+7 mutually adjacent empires with extra adjacencies between empires 2s,5s+2, and 2s+1,5s+3, and 2s+1,5s+1. Thus we obtain a triangular splitting of K_{12s+7} on the plane with excess one.

Case 12s+8. We create an additional empire x_1 by adding triangular components of x_1 at [2s,5s+2,8s+4],[5s+1,8s+3, 11s+5],[8s+2,11s+4,2s], ..., [11s+7,2s+2,5s+4]. Finally we add a single rectangular component of x_1 at [11s+6, 2s+1,5s+3,8s+5] which destroys an extra adjacency between 2s+1 and 5s+3. Thus we obtain a planar map of 12s+8 mutually adjacent empires with two extra adjacencies. From this we obtain an almost perfect splitting of K_{12s+8}

on the plane.

Cases 12s+k for k = 9,10,11,12 are proved similarly completing the proof of Theorem 1. □

In [1] we have constructed perfect splittings of K_n on the torus and Klein bottle when $n \equiv 1 \pmod 6$. Slight modifications of the planar maps we have just constructed can be used to provide almost perfect splittings of K_n on the torus and Klein bottle when $n \equiv 2.5 \pmod 6$ and splittings of excess one when $n \equiv 0,3,4 \pmod 6$. But even more general results of this nature can be obtained.

Each of the maps constructed in proving Theorem 1 have two or three extra adjacencies. To maps of mutually adjacent empires that have extra adjacencies we can sometimes add handles and crosscaps to obtain the following theorem in [4].

Theorem B. For $E(S) \geq -11$, $\sigma(K_n,S) =$

$$\max\{0, \lceil (n-3)(n-4)/6 \rceil - 2+E(S)+e(n)\}$$

where

a) if $n \equiv 0,1 \pmod 3$ then $e(n) = 0$ or 1.

b) if $n \equiv 2 \pmod 3$ then $e(n) = 0$.

When distinct pairs of countries of empires a and b are adjacent we can add a handle as in the top part of Figure 8. The two distinct countries of empire a are merged as well as the two distinct countries of empire b. One extra adjacency between empires a and b is destroyed and one extra adjacency is created. Thus when such a handle is added an almost perfect splitting is transformed to an almost perfect splitting on the new surface and likewise a splitting of excess one is transformed to

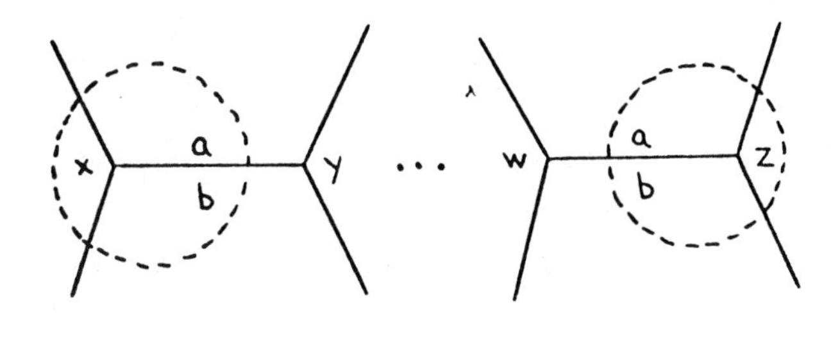

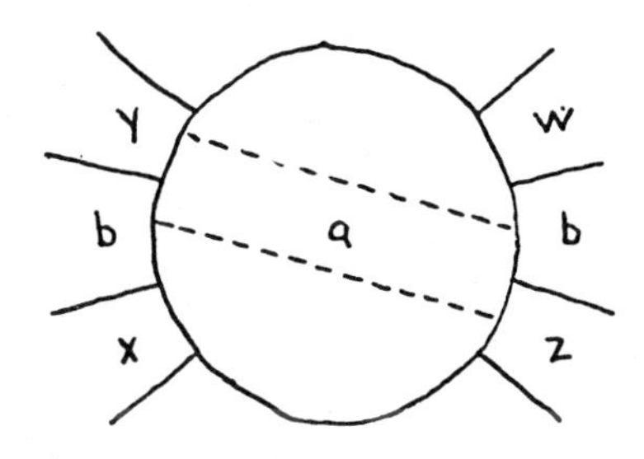

Figure 8

a splitting of excess one on the new surface.

Similarly when distinct countries of empire b are adjacent to a single country of empire a then we can add a crosscap as in the bottom part of Figure 8. Again this operation can be shown to transform an almost perfect splitting to another almost perfect splitting and a splitting of excess one to another splitting of excess one.

We believe that with these techniques one may be able to prove at least some part of the following conjecture.

<u>Conjecture 2</u>. $\sigma(K_n,S)=\max\{0, \lceil(n-3)(n-4)/6\rceil-2+E(S)+e(n)\}$ where

a) if $n \equiv 0,1 \pmod 3$ then $e(n) = 0$ or 1.

b) if $n \equiv 2 \pmod 3$ then $e(n) = 0$.

So far we have been able to show that conjecture 2 is true for $n \leq 13$ and for $E(S) \geq -11$. For example the empire map of 11 mutually adjacent empires in Figure 9 can be used to find splittings of K_{11} on any surface S. It will be much harder to exhibit perfect splittings of K_n when $n \equiv 0,1 \pmod 3$, on an arbitrary surface, but we still conjecture that the following is true.

<u>Conjecture 3</u>. For $n \geq 10$, $\sigma(K_n,S) = \max\{0, \lceil (n-3)(n-4)/6 \rceil - 2 + E(S)\}$.

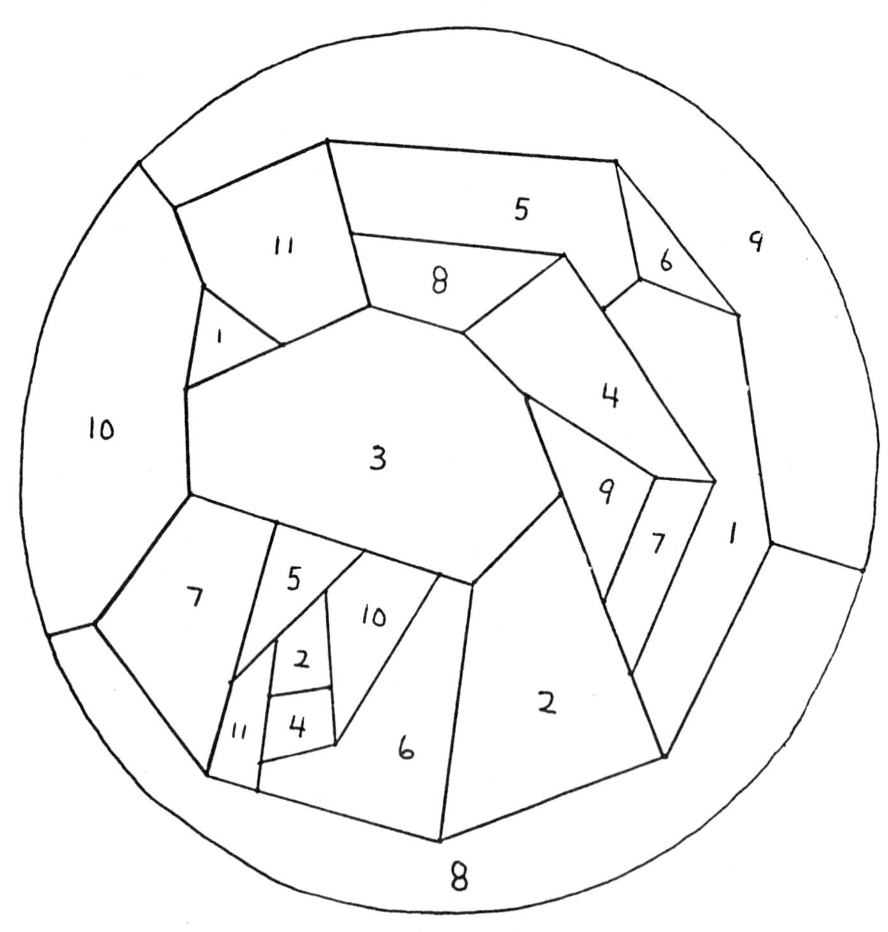

Figure 9

BIBLIOGRAPHY

1. B. Jackson and G. Ringel, Heawood's Empire Problem. J. Combin. Theory, submitted.

2. B. Jackson and G. Ringel, The splitting number of the complete bipartite graph. J. Graph Theory, submitted.

3. B. Jackson and G. Ringel, The planar splitting number of the complete graph, in preparation.

4. B. Jackson and G. Ringel, The splitting number of the complete graph on surfaces, in preparation.

5. B. Jackson and G. Ringel, Empire maps. Proceedings of the First Colorado Symposium on Graph Theory.

MINIMUM INDEPENDENCE GRAPHS
WITH MAXIMUM DEGREE FOUR

Kathryn Fraughnaugh Jones
University of Colorado at Denver
Denver, Colorado 80202

ABSTRACT. Let C be the class of triangle-free graphs with maximum degree at most four. We construct minimum independence graphs in C and derive a lower bound for the number of edges of a graph in C in terms of its independence. Let r(k) be the smallest integer such that every graph in C with r(k) vertices has independence at least k. We evaluate r(k) for all k and obtain 4/13 as the best possible lower bound for the independence ratio for graphs in C.

The relationship between independence of a graph and maximum degree under the further assumption of no large cliques has been investigated by Brooks [2], Albertson, Bollobás and Tucker [1] and Fajtlowicz [3]. A lower bound for the independence ratio in triangle-free graphs with maximum degree three has been found by Staton [8]. The present purpose is to describe a class of minimum independence graphs with maximum degree four and to develop a fairly complete description for maximum degree four.

All graphs under consideration are assumed to be triangle-free and of maximum degree four unless stated otherwise. Following [6], let p, q, and β be the number of vertices, the number of edges, and the size of a maximum independent vertex set of a graph G. If q is the minimum number of edges in any graph with p vertices and independence β, then G is a <u>minimum independence graph</u> or briefly a <u>mingraph</u>. For $k > 0$, the <u>k-chain</u> H_k is a graph of the form of Figure 1 where k is the number of pentagons. For $k > 2$, the <u>extended k-chain</u> E_k is a graph of the form of Figure 2 where k is the number of pentagons. For $k > 2$, D_k is a graph of the form of Figure 3 where k is the number of pentagons in the extended k-chain which occurs as a subgraph.

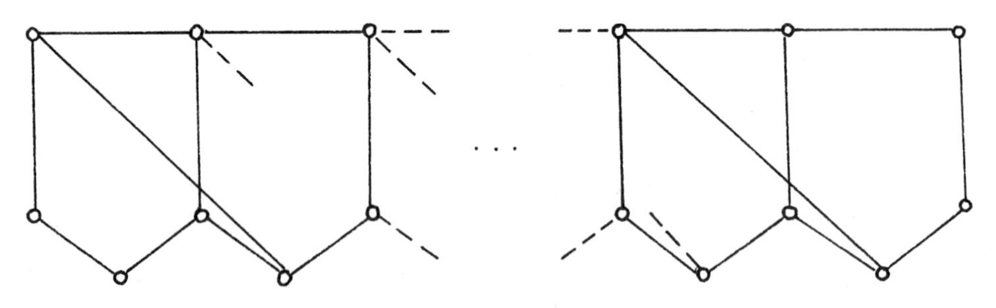

<u>Figure 1</u>. The k-chain H_k.

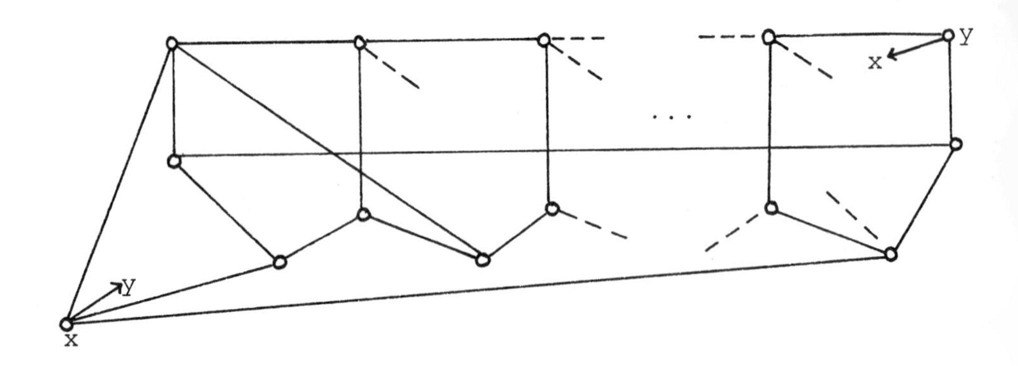

<u>Figure 2</u>. The extended k-chain E_k.

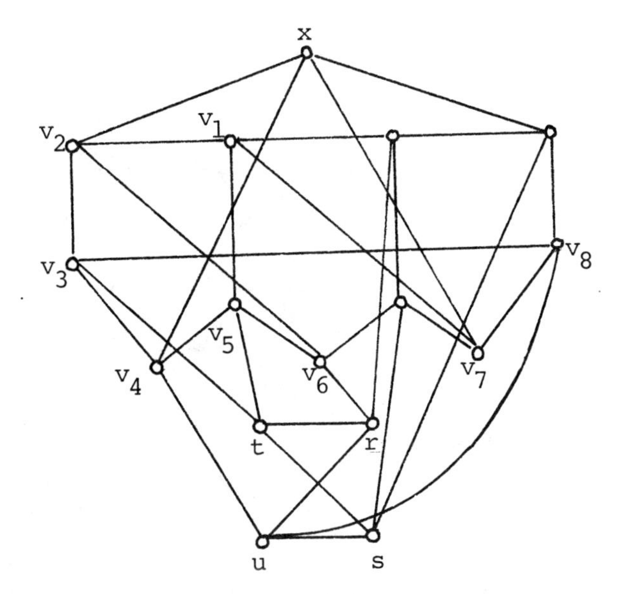

Figure 3(a). D_3

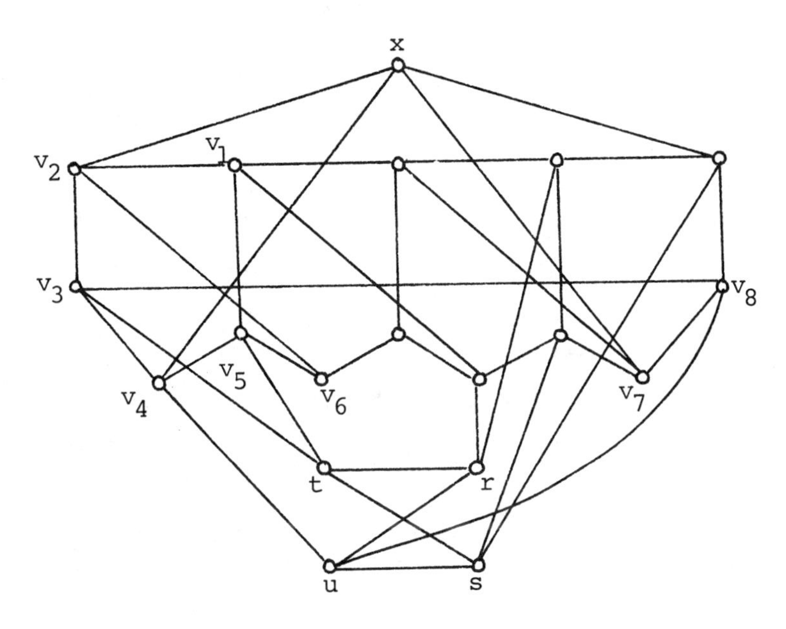

Figure 3(b). D_4

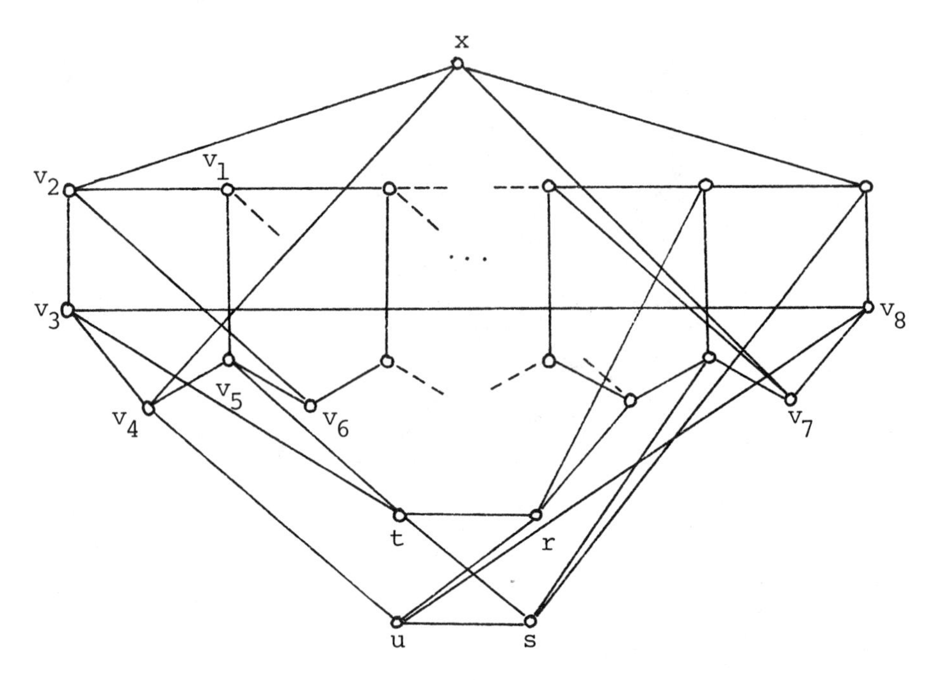

Figure 3(c). D_k for $k > 4$.

In the next three propositions, we find the values of p, q, and β for the graphs H_k, E_k, and D_k.

Proposition 1. If G is the k-chain H_k for some $k > 0$, then $p = 3k + 2$, $q = 5k$, and $\beta = k + 1$.

Proof. The only statement which is not obvious from the construction is that $\beta = k + 1$. For $k = 1$, the statement is trival since H_1 is the pentagon. When $k = 2$, we have the graph H_2 with independence 3 found by Graver and Yackel [5]. Assume the statement for $k \leq n$ and let $G = H_{n+1}$. We note that the deletion of the three vertices which belong only to the first pentagon leaves the graph H_n which has independnece $n + 1$ by induction. We can construct an independent set of $n + 2$ vertices by choosing

$n+1$ independent vertices from this H_n and the degree-2 vertex in the first pentagon which is adjacent to no vertex in the remaining H_n. This establishes that $\beta \geq n+2$. Since $\beta(H_n) = n+1$, an independent set with $n+3$ vertices must contain the two independent vertices in the first pentagon. However, if we delete the closed neighborhoods of both these vertices, the graph which remains is H_{n-1}, which by induction has independence n. Thus an independent set containing these two vertices has as most $n+2$ vertices and $\beta = n+2$. □

Proposition 1 yields some additional information about independence in planar graphs. It follows from Grötsch's Theorem [4] that, for a planar triangle-free graph G with arbitrary maximum degree, the <u>independence ratio</u> β/p is at least 1/3. When G has maximum degree at most three, Staton [7] proved that this ratio can be improved to 5/14 without the assumption of planarity. However, since a k-chain is a planar graph whose independence ratio approaches 1/3 as k becomes large, no improvement is possible for graphs with maximum degree four.

Proposition 2. If G is E_k for some $k > 2$, then $p = 3(k+1)$, $q = 5(k+1)$ and $\beta = k+1$.

<u>Proof</u>. Again we will prove only the statement about independence. We observe that E_k is obtained from H_k by the addition of the vertex x'; see Figure 2. By Proposition 1, $\beta(H_k) = k+1$, so that $\beta \geq k+1$ and if there is an independent set in E_k with $k+2$ vertices, then it must contain x. The deletion of the closed neighborhood of x leaves $H_{k-2} \cup K_2$. Since $\beta(H_{k-2} \cup K_2) = \beta(H_{k-2}) + \beta(K_2) = k$, every independent set containing x has independence at most $k+1$. □

Proposition 3. If G is D_k for some $k > 2$, then $p = 3(k+2) + 1$, $q = 5(k+3) + 2$ and $\beta = k+2$.

Proof. We will prove $\beta = k + 2$. Since by Proposition 2, $\beta(E_k) = k + 1$, if there is an independent set S with $k + 3$ vertices, then S must contain an independent pair of vertices from the 4-cycle $D_k - E_k$. If we label the vertices as in Figure 3, then either t and u belong to S or r and s belong to S.

Suppose t and u belong to S and consider the graph X which is obtained by deleting the closed neighborhoods of both t and u. The mapping which takes x to v_8, v_2 to v_5, and leaves all other vertices of X fixed is an isomorphism between X and H_{k-1} with an additional edge from x to v_2. Since $\beta(H_{k-1}) = k$ by Proposition 1, we have $|S| \leq k + 2$. Notice that we can construct an independent set with $k+2$ vertices containing t and u and a k-element independent set from H_{k-1}, which establishes that $\beta \geq k + 2$.

Now suppose r and s belong to S and consider the graph Y which is obtained by deletion of their closed neighborhoods. We observe that v_2, x, v_7, v_8, and v_3 form a pentagon. If we consider this as the first pentagon in a chain and observe that x is adjacent to v_4, we see that Y is isomorphic to H_{k-1} with an additional edge incident to v_7. Again by Proposition 1, $\beta(H_{k-1}) = k$ and $|S| \leq k + 2$. □

The next two propositions summarize some properties of E_k.

Proposition 4. for $k \geq 3$, the graph E_k contains $2(k+1)$ vertices with degree 3, each of which has exactly one neighbor of degree 4. The vertices of E_k with degree 3 form a cycle.

Proof. The number of vertices of degree 3 in a graph with maximum degree 4 is $2(2p - q)$. It follows from this fact and Proposition 2 that the number of vertices of degree 3 in E_k is $2(k+1)$. The other statements follow from the construction of E_k. □

Proposition 5. Let G be a graph which contains an induced subgraph H isomorphic to E_k for some $k \geq 3$ and let q' be the number of edges with exactly one endpoint in H. If $1 \leq q' < 2(k + 1)$, then there is a vertex in H whose degree in both H and G is 3 and which has at least two neighbors of degree four.

Proof. Let K be the set of vertices of H which are incident to an edge with exactly one endpoint in H. Then $q' = |K|$, so that $q' \geq 1$ implies that K is nonempty. Since G has maximum degree 4 and E_k has minimum degree 3, each vertex in K has degree 3 in H and degree 4 in G. Thus we have $q' \leq 2(k+1)$, which is the number of vertices with degree 3 in H by Proposition 4. Since $q' < 2(k + 1)$, there is a vertex x of degree 3 in H such that $x \notin K$ and consequently x has degree 3 in G. Moreover, since by Proposition 4 the vertices of degree 3 constitute a cycle, x may be chosen adjacent to a vertex $y \in K$. Recall that since $y \in K$, y has degree 3 in H and degree 4 in G. But by Proposition 4, x already has a neighbor in H with degree 4 in H, so that x has at least two neighbors of degree four. □

The following theorem provides a relationship between the number of edges in a graph and its independence and characterizes certain minimum independence graphs.

Theorem 1. If G is a graph, then $q \geq 6p - 13\beta$ and if equality holds, then each component of G is either 4-regular or an extended k-chain.

Outline of proof. The proof involves characterization not only of graphs with $6p - 13\beta$ edges but also of graphs with $6p - 13\beta + 1$ edges. The idea of the proof is to consider a graph G and to delete from G certain crucial arrays of vertices and edges. The deletion is chosen in such a way that the subgraph which remains, by induction, has components which are 4-regular or chains or else contain such graphs as large subgraphs. The properties of chains lead to contradiction in case the edge inequality is violated. The full proof is quite long and tedious and involves several technical lemmas. Further details may be found in [7].

The graphs which we have constructed are of additional interest in that they provide examples of minimum independece graphs. We observe that if G is an extended k-chain, then $q = 6p - 13\beta$, so that E_k is a mingraph for all $k > 2$. On the other hand, if G is H_k or D_k, then $q = 6p - 13\beta + 1$. Nevertheless, it has been shown [7] that, for $1 \leq k \leq 6$, H_k is a mingraph and, for $1 \leq k \leq 5$, it is unique with respect to this property. Similarly, for $3 \leq k \leq 5$, D_k is a mingraph [7]. In addition, it is known [5] that H_1, H_2, H_3, E_3 and E_4 are the unique mingraphs in the class of triangle-free graphs with arbitrary maximum degree. The graph D_4 provides insight into a question considered by Graver and Yackel [5]. They showed that the minimum number of edges possible in a graph with 19 vertices and independence 6 is either 36 or 37. Since D_4 is a mingraph, the smaller value can only be attained in a graph with a vertex of degree 5 or more.

We now state and prove two important corollaries to the theorem.

Corollary 1. The best possible lower bound for the independence ratio β/p of a graph G is $4/13$.

Proof. If G is a graph with maximum degree four, then $q \leq 2p$. It follows from the theorem that $6p - 13\beta \leq 2p$. Solving for β/p, we obtain $\beta/p \geq 4/13$ as asserted. The well-known graph on 13 vertices with independence 4 found in [5] is an example where the ratio is as small as possible. □

Let $r(k)$ be the smallest positive integer such that any graph G with $p \geq r(k)$ has $\beta \geq k$. That such an integer exists follows from Brooks' Theorem [2]. We note that $r(k)$ coincides with the classical Ramsey number for $k \leq 5$. A graph is critical if it has $r(k) - 1$ vertices and independence $\leq k - 1$. Let R be the graph on 13 vertices with independence 4 found in [5]. The graph which consists of m disjoint copies of R will be denoted mR as usual.

Corollary 2. For each positive integer n and for $s = 1$, 2, 3, 4, the value of $r(k)$ where $k = 4n + s$ is given by $r(4n + s) = 13n + 3s - 2$.

Proof. Let G be a graph with $p = 13n + 3s - 2$. It follows from Corollary 1 that $\beta \geq (4/13)p = (4/13)(13n + 3s - 2) = 4n + s - (s + 8)/13$. But since $s \leq 4$, we have $(s + 8)/13 < 1$ and since β is an integer, we have $\beta \geq 4n + s$. With this, we have shown that $r(4n + s) \leq 13n + 3s - 2$. In order to show equality, we must display a critical example. For $s = 1$, a critical example is nR

for which $p = 13n$ and $\beta = 4n$. For $s = 2, 3$, or 4, we consider the graph $(n-1)R \cup D_{s+1}$. It follows from Proposition 3 that this graph has $p = 13(n-1) + 3(s+3)+1 = 13n + 3s - 3$ and $\beta = 4(n-1) + s + 3 = 4n + s - 1$. □

REFERENCES

1. Michael Alberton, Bela Bollobás, and Susan Tucker, The independence ratio and maximum degree of a graph. Congressus Numerantium 17 (1976), 43-50.
2. R. L. Brooks, On colouring the nodes of a network. Proc. Cambridge Phil. Soc. 37 (1941), 194-197.
3. S. Fajtlowicz, On the size of independent sets in graphs. Congressus Numerantium 21 (1978), 269-274.
4. B. Grünbaum, Grötsch's theorem on 3-colorings. Michigan Math J. 10 (1963), 303-310.
5. Jack E. Graver and James Yackel, Some graph theoretic results associated with Ramsey's theorem. J. Combin. Theory 4 (1968), 125-175.
6. F. Harary, Graph Theory, Addison-Wesley, Reading (1969).
7. K. Jones, Independence in graphs with maximum degree four. Doctoral dissertation, University of Houston (1982).
8. W. Staton, Some Ramsey-type numbers and the independence ratio. Trans. of A.M.S. 256 (1979), 353-370.

BLOCK GRAPHS AND THEIR MATRIX DUALS

J. Richard Lundgren
University of Colorado at Denver
Denver, Colorado 80202

John S. Maybee
University of Colorado at Boulder
Boulder, Colorado 80309

ABSTRACT. The class of block graphs introduced by Harary is studied with the help of certain rectangular matrices. A concept of natural duality is introduced and we prove that all of the natural duals of a block graph G are also block graphs. We define maximum and minimum duals of a block graph G and characterize min max G and max min G. The special cases where G is a simple block graph (each cutpoint belongs to exactly two blocks) and a tree are also studied. We show that there are dual theories for these two classes.

1. INTRODUCTION

We will develop a special kind of duality theory for the class of graphs called block graphs (see Harary [3]). A graph G is called a block graph if it is the block graph of some graph H. Equivalently G is a block graph if it is a connected graph in which each block is complete.

Our duality theory is based upon the relation between rectangular matrices and graphs established in the papers [1] and [2]. Let us briefly review these ideas.

Let A be an $m \times n$ boolean matrix (elements 0 and 1) and denote by $C = \{c_1, c_2, \ldots, c_n\}$ the set of columns of A and by $R = \{r_1, r_2, \ldots, r_m\}$ the set of rows of A. Then the <u>column graph</u> CG(A) and the <u>row graph</u> RG(A) are defined as follows. For CG(A) the set of points is the set C and there is a line joining c_i and c_j if there exists at least one k such that $a_{ki} \neq 0$ and $a_{kj} \neq 0$. Similarly the set of points of RG(A) is the set R and a line joins r_i and r_j if there exists at least one k such that $a_{ik} \neq 0$ and $a_{jk} \neq 0$.

A boolean matrix A is called a <u>column inverse</u> of the graph G if $G \cong CG(A)$. Given any graph G there are in general many column inverses of G as was shown in [2]. These may be obtained in the following way. Define a <u>clique cover</u> of the graph G to be a finite set of cliques of G such that every line and every point of G belongs to at least one clique. As an example consider the block graph shown in Figure 1.

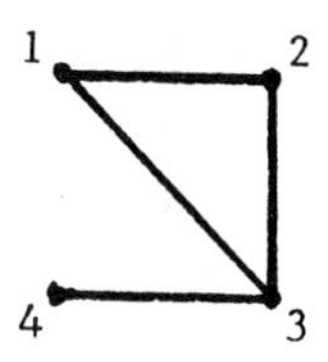

<u>Figure 1</u>. The smallest block graph with different blocks

The following sets are clique covers of G :

$$S_1 = \{\langle 1,2,3\rangle, \langle 2,4\rangle, \langle 2\rangle\},$$
$$S_2 = \{\langle 1,2\rangle, \langle 2,3\rangle, \langle 3,4\rangle, \langle 1,2,3\rangle\},$$
$$S_3 = \{\langle 1,2,3\rangle, \langle 3,4\rangle\},$$
$$S_4 = \{\langle 1,2\rangle, \langle 1,3\rangle, \langle 3,4\rangle, \langle 3,4\rangle, \langle 2,3\rangle\},$$

$S_5 = \{\langle 1,2,3\rangle, \langle 3,4\rangle, \langle 1\rangle, \langle 2\rangle, \langle 4\rangle\}$.

Here we have used $\langle X\rangle$ to denote the subgraph of G generated by the points in the set X.

Now matrices A_i such that $CG(A_i) \cong G$ can be constructed from the clique covers S_i by assigning a row of the matrix to each clique and placing 1's in the designated columns. Thus to S_1 through S_5 above there correspond the matrices

$$A_1 = \begin{bmatrix} 1 & 1 & 1 & 0 \\ 0 & 0 & 1 & 1 \\ 0 & 1 & 0 & 0 \end{bmatrix} \quad A_2 = \begin{bmatrix} 1 & 1 & 0 & 0 \\ 0 & 1 & 1 & 0 \\ 0 & 0 & 1 & 1 \\ 1 & 1 & 1 & 0 \end{bmatrix} \quad A_3 = \begin{bmatrix} 1 & 1 & 1 & 0 \\ 0 & 0 & 1 & 1 \end{bmatrix},$$

$$A_4 = \begin{bmatrix} 1 & 1 & 0 & 0 \\ 1 & 0 & 1 & 0 \\ 0 & 0 & 1 & 1 \\ 0 & 0 & 1 & 1 \\ 0 & 1 & 1 & 0 \end{bmatrix} \quad A_5 = \begin{bmatrix} 1 & 1 & 1 & 0 \\ 0 & 0 & 1 & 1 \\ 1 & 0 & 0 & 0 \\ 0 & 1 & 0 & 0 \\ 0 & 0 & 0 & 1 \end{bmatrix}$$

2. NATURAL DUALS

For block graphs we can define a set of column inverses that have particularly nice properties. We do this in the following way. If G is a block graph a clique cover of G will be called natural if it consists of the set of blocks of G together with a subset of the set of noncutpoints of G. We do not allow a clique to be repeated in a natural clique cover. In the example above S_3 and S_5 are the only natural clique covers. The maximum natural clique cover of G, S(G), contains all noncutpoints of G and the minimum natural clique cover of G, s(G), contains none of the noncutpoints of G. Again, in the example $s(G) = S_3$ and $S(G) = S_5$.

If the matrix A corresponds to a natural clique

cover of G we call it a natural column inverse of G. We will use the notations $\bar{A}$ to denote the natural column inverse of G corresponding to S(G) and $\underline{A}$ to denote the natural column inverse of G corresponding to s(G). Note that a column inverse of a graph G is uniquely determined by a clique up to the order of its rows and columns.

Now the graph H is called a natural dual of the graph G if H ≅ RG(A) for some natural column inverse A of G.

Our main result is the following.

Theorem 1. If G is a block graph then every natural dual of G is also a block graph.

Proof. Let A_N be a natural column inverse of G. Without loss of generality we may suppose that the first p rows of A_N correspond to the blocks of G and the remaining rows to the noncutpoints (if any). Thus if G has n points we have

$$A_N = \begin{bmatrix} A_1 \\ A_2 \end{bmatrix}$$

where A_1 is a $p \times n$ block and A_2 a $q \times n$ block. Each column of A_1 has a simple nonzero element unless it corresponds to a cutpoint of G, in which case it has S ones where S is the number of blocks of G to which the cutpoint belongs. Each row of A_2 has a single nonzero entry so that A_2 has q columns with exactly one nonzero and n-q columns which are zero. The columns of A_2 having a nonzero do not correspond to columns of A_1 having more than one nonzero. Thus RG(A_N) has q 2-cliques and its remaining cliques correspond to cutpoints of G. We claim first that no two cliques of RG(A_N) intersect at more than one point. This is obvious for pairs of cliques where at least one is a 2-clique corresponding to a row of A_2. Thus consider two cliques corresponding to cutpoints of G. For them to have at least two points in common, A_1 must have a 2×2 submatrix of the form $\begin{bmatrix} 1 & 1 \\ 1 & 1 \end{bmatrix}$. But this means that two blocks of G have a line in common, contradicting the fact that G is a block graph. Finally we claim that $RG(A_N)$ cannot have a chain of cliques $C_1, C_2, \ldots, C_k$ with $C_1 \cap C_2 \neq \phi, \ldots, C_{k-1} \cap C_k \neq \phi$, $C_k \cap C_1 \neq \phi$ and $C_i \cap C_j = \phi$ for $j \neq (i+1) \bmod k$. Clearly the 2-cliques formed from elements in A_2 cannot form part of such a chain because they are only connected to one other clique of $RG(A_N)$. Thus if such a chain exists, then A_1 must have an $k \times k$ submatrix which, after suitable column permutation, has the form

$$A_{10} = \begin{bmatrix} 1 & 0 & 0 & . & . & . & 0 & 1 \\ 1 & 1 & 0 & & & & & \\ 0 & 1 & 1 & & & & & \\ . & . & . & & & & & \\ 0 & 0 & 0 & . & . & . & 1 & 0 \\ 0 & 0 & 0 & . & . & . & 1 & 1 \end{bmatrix}$$

in which A_{10} has 1's on the principal diagonal and the first subdiagonal and zero's above the principal diagonal except for a 1 in row 1 column k. But this implies that G has a chain of cliques, contradicting the fact that G is a block graph. Finally, [1, Theorem 1], $RG(A_N)$ is connected. It follows that $H = RG(A_N)$ is a block graph. □

Remark 1. Note that if A is not a natural column inverse of G, then RG(A) need not be a block graph. For example, the row graph of the matrix A_2 above is shown in Figure 2. It is clearly not a block graph.

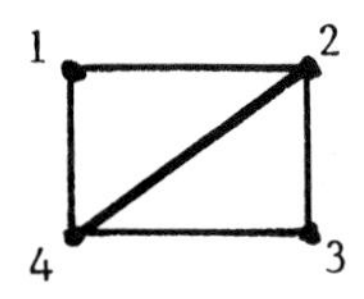

Figure 2. The random graph.

Remark 2. Our proof of Theorem 1 sheds light upon the connection between $H = RG(A_N)$ and G, when H is a natural dual of G. The blocks of H correspond to the cutpoints of G augmented by blocks which are 2-cliques corresponding to the subset of noncutpoints used to form A_N. On the other hand, the cutpoints of H correspond to certain cliques of G. A block of G corresponds to a cutpoint of H if it contains more than one cutpoint of G or if it contains only one cutpoint and a noncutpoint used in forming A_N.

Remark 3. A question that arises is the following. Suppose H is a natural dual of G. Is G then a natural dual of H? The answer is no. To see this consider the block graph G of Figure 1. A natural column inverse is the matrix

$$A = \begin{bmatrix} 1 & 1 & 1 & 0 \\ 0 & 0 & 1 & 1 \\ 0 & 1 & 0 & 0 \end{bmatrix}$$

giving rise to the dual H shown in Figure 3. For a

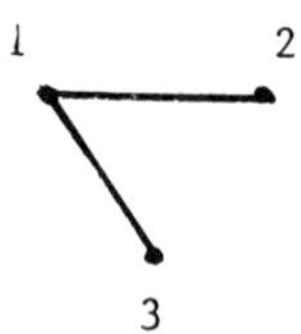

Figure 3. A dual of G.

natural dual of H to be G, the natural column inverse of H must have four rows. Thus it must be the matrix A'

$$A' = \begin{bmatrix} 1 & 1 & 0 \\ 1 & 0 & 1 \\ 0 & 1 & 0 \\ 0 & 0 & 1 \end{bmatrix}$$

which does not have G as its row graph. Note that the transpose matrix

$$A^T = \begin{bmatrix} 1 & 0 & 0 \\ 1 & 0 & 1 \\ 1 & 1 & 0 \\ 0 & 1 & 0 \end{bmatrix}$$

does have G as its row graph as is required by [2, Theorem 1], but it is not a natural column inverse of H.

3. MAXIMUM AND MINIMUM DUALS

The most interesting natural duals of G are those obtained from $RG(\bar{A})$ and $RG(\underline{A})$.

Notation. Let us set $RG(\bar{A}) = \max G$ and $RG(\underline{A}) = \min G$ for G a block graph. We start with the graph max G which has the following interesting property.

Theorem 2. If G is a block graph, then min max G = G.

Proof. Consider the matrix $\bar{A}$. If G has p blocks we can assume without loss of generality that the first p rows of $\bar{A}$ correspond to the blocks of G and the remaining rows are single rows, one for each noncutpoint of G. From the proof of Theorem 1 we can conclude that each column of $\bar{A}$ corresponds to a block of max G. It follows that the matrix $\bar{A}^T$ is a natural column inverse of max G and, since it has no singleton rows, must therefore be $\underline{A}(\max G)$. But $RG(\bar{A}^T) \cong G$, hence G = min max G as was to be shown. ▯

As an example consider the graph of Figure 4.

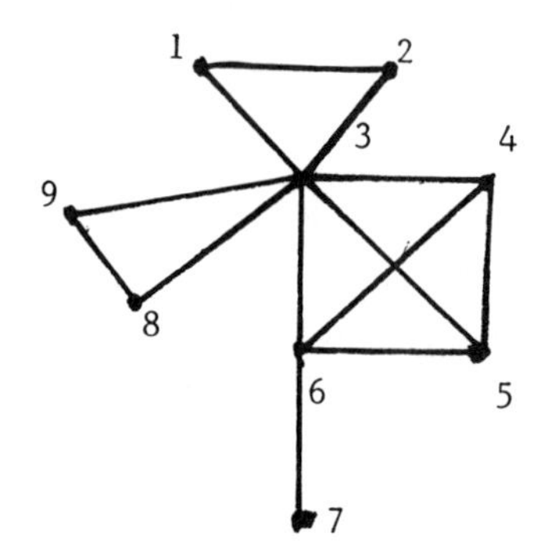

Figure 4. Graph of $\bar{A}$.

$$\bar{A} = \begin{bmatrix} 1 & 1 & 1 & 0 & 0 & 0 & 0 & 0 & 0 \\ 0 & 0 & 1 & 1 & 1 & 1 & 0 & 0 & 0 \\ 0 & 0 & 0 & 0 & 0 & 1 & 1 & 0 & 0 \\ 0 & 0 & 1 & 0 & 0 & 0 & 0 & 1 & 1 \\ 1 & 0 & 0 & 0 & 0 & 0 & 0 & 0 & 0 \\ 0 & 1 & 0 & 0 & 0 & 0 & 0 & 0 & 0 \\ 0 & 0 & 0 & 1 & 0 & 0 & 0 & 0 & 0 \\ 0 & 0 & 0 & 0 & 1 & 0 & 0 & 0 & 0 \\ 0 & 0 & 0 & 0 & 0 & 0 & 1 & 0 & 0 \\ 0 & 0 & 0 & 0 & 0 & 0 & 0 & 1 & 0 \\ 0 & 0 & 0 & 0 & 0 & 0 & 0 & 0 & 1 \end{bmatrix}$$

The graphs max G and min G are shown in Figure 5.

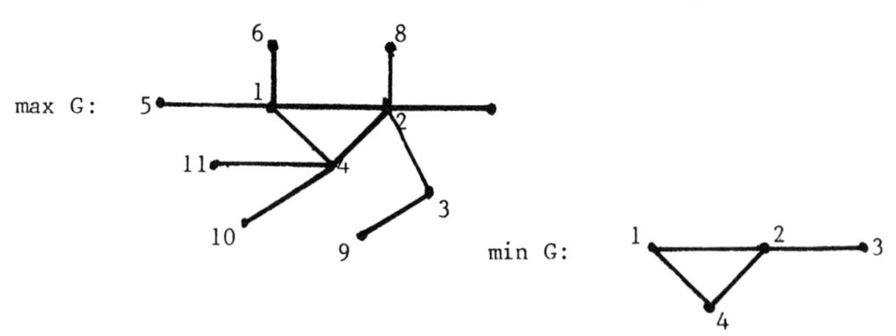

Figure 5. Maximum and minimum duals

In order to obtain a result similar to Theorem 2 for the graph min G, we must introduce another concept. Let x, y, z be distinct points of G such that y is not a cutpoint of G and [x,y], [y,z], [x,z] are all lines of G. Then we call the graph G-y a point retraction of G. Let G_0 be the graph obtained by performing all possible successive point retractions of G. Given G, the graph G_0 is uniquely determined from G. For example the graph G_0, associated with the graph G of Figure 4 is shown in Figure 6. We have chosen to remove the points 2,8,4,5,

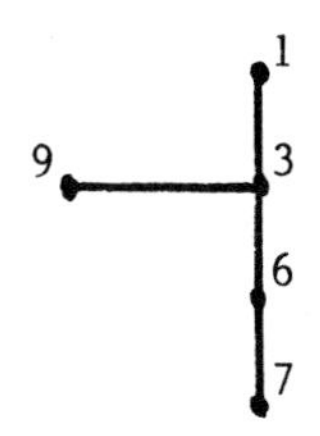

Figure 6. G_0 associated with G.

but any other sequence or retractions would lead to a G_0 isomorphic to the graph shown. The graph G_0 has the property that it is the graph with the smallest number of points having the same block cutpoint graph as G. We can now prove our next result.

Theorem 3. If G is a glock graph, then max min G = G_0.

Proof. Consider the matrix $\underline{A}$. Each row of $\underline{A}$ corresponds to a block of G and those columns of $\underline{A}$ corresponding to cutpoints of G correspond to the blocks of min G. Now if G is not a tree the matrix $\underline{A}^T$ is not a natural column inverse of min G because it has repeated singleton rows and singleton rows corresponding to cutpoints of min G. However, let A_0 be the matrix obtained from $\underline{A}^T$ by deleting repeated singleton rows until there are no repetitions and by deleting all singleton rows corresponding to cutpoints of min G. Since G = RC($\underline{A}^T$)

deleting each such singleton row of $\underline{A}^T$ is equivalent to making a retraction of G. Thus the deletion of all such singleton rows is equivalent to producing a matrix A_0 whose row graph is isometric to G_0. On the other hand, when all the rows have been deleted we are left with a matrix whose nonsingleton rows correspond to the blocks of min G and with a singleton row for each noncutpoint of min G. Thus $A_0 = \bar{A}(\min G)$. It follows that $G_0 = \max \min G$. □

For the graph of Figure 4 we have

$$\underline{A} = \begin{bmatrix} 1 & 1 & 1 & 0 & 0 & 0 & 0 & 0 & 0 \\ 0 & 0 & 1 & 1 & 1 & 1 & 0 & 0 & 0 \\ 0 & 0 & 0 & 0 & 0 & 1 & 1 & 0 & 0 \\ 0 & 0 & 1 & 0 & 0 & 0 & 0 & 1 & 1 \end{bmatrix} \qquad A^T = \begin{bmatrix} 1 & 0 & 0 & 0 \\ 1 & 0 & 0 & 0 \\ 1 & 1 & 0 & 1 \\ 0 & 1 & 0 & 0 \\ 0 & 1 & 0 & 0 \\ 0 & 1 & 1 & 0 \\ 0 & 0 & 0 & 1 \\ 0 & 0 & 0 & 1 \\ 0 & 0 & 0 & 1 \end{bmatrix}$$

If we obtain A_0 from $\underline{A}^T$ by crossing out the singleton rows 2,4,5,8 then A_0 has the tree of Figure 6 as its row graph.

It is useful to examine the set of natural duals of a block graph G in terms of min G. We have seen that min G has a block corresponding to each cutpoint of G and that the block is the complete graph K_p if the cutpoint belongs to p blocks $(p > 2)$ of G. Obviously, min G is a subgraph of H if H is any natural dual of G. In fact, each such H is obtained from min G by attaching lines to certain points of min G. Let us call the noncutpoint x of min G a <u>critical point</u> if for at least one natural dual H of G, x is a cutpoint of H. How do we recognize critical points of min G?

Consider the natural column inverse $\underline{A}$. The points of min G correspond to the rows of $\underline{A}$ so, if row i has nonzero elements in columns $j_1, \ldots, j_r$, consider the submatrix of $\underline{A}$ obtained by striking out all columns of $\underline{A}$ having zero elements in row i and by striking out row i. If $\underline{A}$ is an $n \times q$ matrix, the resulting matrix $\underline{A}_i$ is an $r \times (q-1)$ matrix. The point i in RG($\underline{A}$) is a critical point if A_i has at least one column of zeros.

As an example consider the graph of Figure 4 whose natural column inverse $\underline{A}$ is displayed above. It is easy to see that $\underline{A}_i$ has a column of zeros for $1 \leq i \leq 4$.

Since the points of min G correspond to the blocks of G, we see that a point in min G is a critical point if and only if it corresponds to a block of G containing at least one noncutpoint.

4. SIMPLE BLOCK GRAPHS

We will call the block graph G simple if each cutpoint of G belongs to exactly two blocks.

Theorem 4. Let G be a block graph. Then G is simple if and only if every natural dual of G is a tree.

Proof. To prove the only if portion, suppose G is a simple block graph and let A_N be a natural column inverse of G. Then, as in the proof of Theorem 1, we can write

$$A_N = \begin{bmatrix} A_1 \\ A_2 \end{bmatrix}$$

where A_1 has p rows if G has p blocks. Since the

blocks of $RG(A_N)$ correspond to the cutpoints of G and each cutpoint belongs to exactly two blocks, no column of A_N contains more than two elements. Thus, if $H = RG(A_N)$, every block of H is a line of H. Finally, by [1, Theorem 1], it is connected because G is connected. Thus H is a tree.

For the converse suppose each natural dual H of G is a tree. Since $H = RG(A_N)$ for some A_N and H is a tree, no column of A_N can have more than two nonzero elements. It follows that every cutpoint of G belongs to exactly two blocks of G. Thus G is a simple block graph. ▯

The block graph shown in Figure 7 is simple.

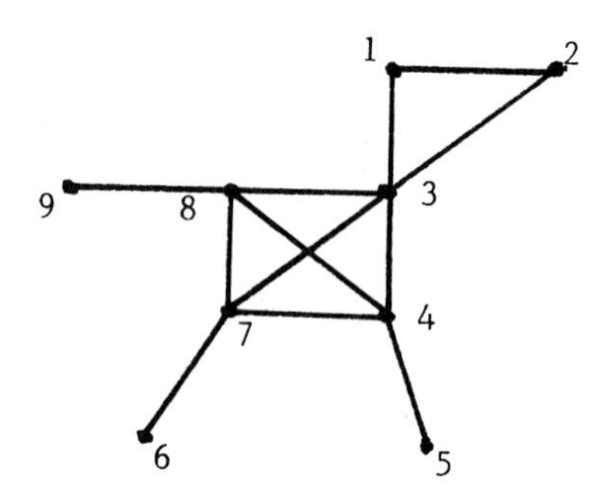

Figure 7. A simple block graph

Three natural duals are illustrated in Figure 8. Figure 8a is min G, Figure 8c is max G, and Figure 8b is another dual.

Here is a result concerning trees which complements Theorem 4.

Theorem 5. Let G be a block graph. Then G is a tree if and only if every natural dual of G is a simple block graph.

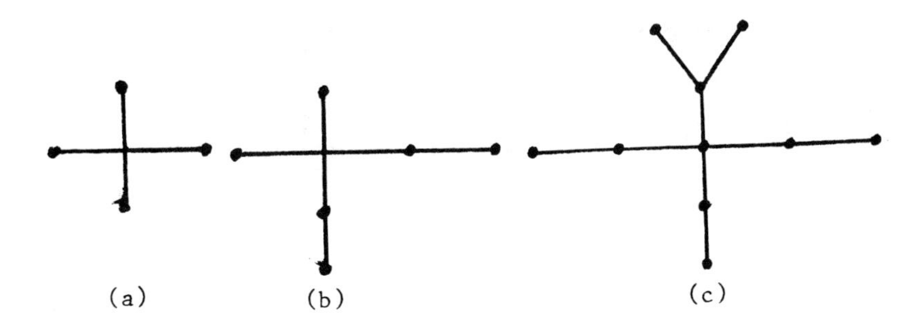

Figure 8. Some natural duals.

Proof. Suppose G is a tree on n points. Then G is a block graph with n-1 blocks. It follows that any natural column inverse A_N of G can be written in the form

$$A_N = \begin{bmatrix} A_1 \\ A_2 \end{bmatrix}$$

where A_1 has n-1 rows, each one of which has exactly two nonzero elements and A_2 consists of singleton rows. But the fact that no row of A_N contains more than two nonzero elements means that no point of $RG(A_N)$ belongs to more than two blocks, hence $H = RG(A_N)$ is a simple block graph. To show that, if all natural duals of G are simple block graphs, then G is a tree, we need only look at $\bar{A}$. Every column of $\bar{A}$ has at least two nonzero elements, hence every column corresponds to a block of max G. But this implies that no row of A_N can have more than two points because max G is a simple block graph. Thus the blocks of G are lines and, since G is connected, G must be a tree. ▯

The tree shown in Figure 8b has the minimum and maximum duals shown in Figure 9a and 9b.

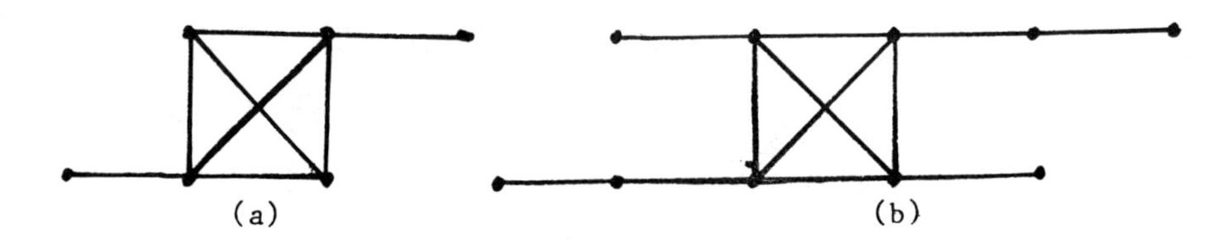

Figure 9. Minimum and maximum duals.

REFERENCES

1. H. Greenberg, R. Lundgren and J.S. Maybee, Graph theoretic methods for the qualitative analysis of rectangular matrices, SIAM J. Alg. Disc. Meth. 2, (1981), 227-239.
2. H. Greenberg, R. Lundgren and J.S. Maybee, Inverting graphs of rectangular matrices (to appear).
3. F. Harary, Graph Theory, Addison-Wesley, Reading (1969).

FOOD WEBS WITH INTERVAL COMPETITION GRAPH

J. Richard Lundgren
University of Colorado at Denver
Denver, CO 80202

John S. Maybee
University of Colorado
Boulder, CO 80309

ABSTRACT. The problem of determining the structure of a digraph and its adjacency matrix when the competition graph of the digraph is an interval graph is investigated. Sufficient conditions on the structure of the matrix, and necessary and sufficient conditions on the structure of the digraph are given. The dual problem for common enemy graphs is also investigated, and a characterization given in terms of source induced subdigraphs. Finally, a possible method is suggested for trying to find a forbidden sink induced subdigraph list for digraphs with interval competition graphs.

In 1968 Cohen [1] introduced competition graphs associated with food web models of an ecosystem as a means of determining the dimension of ecological phase space. Surprisingly, all initial food webs studied by Cohen have competition graphs that are interval graphs. However, further study by Cohen [2] produced examples of food webs whose competition graphs are not interval. Nevertheless, most competition graphs arising from food webs are interval graphs, particularly those which describe single habitats. Hence, an important ecological question is to characterize those food webs whose competition graphs are interval.

Let D be a digraph. The competition graph of D is the undirected graph G obtained as follows: G has the same vertices as D and {x,y} is an edge in G if and only if for some vertex z in D there are arcs (x,z) and (y,z) in D. If we let A be the adjacency matrix of D, then we see that G is simply the "row graph" of A. In the row graph G = RG(A) studied by Greenberg, Lundgren and Maybee [5,6], the rows of A are the points of G, and two rows are adjacent in G if and only if they have a nonzero entry in the same column of A. It is interesting to note that the column graph CG(A), defined similarly, can also be given an ecological interpretation. As two points are adjacent in CG(A) if and only if they have a common predator in D, we also call CG(A) the common enemy graph.

Ecologists usually assume that the digraph D is acyclic. Competition graphs of acyclic digraphs have been characterized by Dutton and Brigham [3] and Lundgren and Maybee [9]. Competition graphs of digraphs D which may not be acyclic have been characterized by Dutton and Brigham [3] and Roberts and Steif [16]. Several other questions about this general problem have been investigated by Opsut [13], Roberts [14, 15] and Steif [17].

We now turn to several questions related to the question about interval graphs raised above. What (acyclic) digraphs have competition graphs that are interval?

If the competition graph is an interval graph, is the common enemy graph an interval graph? Is there a special significance to the case where both graphs are interval graphs? Can we determine if RG(A) is an interval graph from the structure of A? Does there exist a forbidden subdigraph characterization of those digraphs whose competition graphs are interval (see Steif [17])? We will provide answers to some of these questions.

In what follows we assume that D is a digraph representing a food web, A is the adjacency matrix of D, and G = RG(A) is the competition graph. First observe that the columns of A either determine a clique in G or consist entirely of zeros (see [6] for details). We then get the following result which is similar to the Fulkerson-Gross theorem on the vertex maximal cliques incidence matrix [4].

Theorem 1. If the rows of A have the consecutive ones property, then G = RG(A) is an interval graph.

Proof. Let $C_1,\ldots,C_n$ be the cliques of G = (V,E) corresponding to the columns of A. If $v \in V$ and v is in one of these cliques, let J(v) = [i,j] where i is the least integer such that $v \in C_i$ and j is the greatest integer such that $v \in C_j$. For each of the points v not in any of these cliques we assign a different interval of the type [k,k] with k > n. Since the rows of A have the consecutive ones property, it is easy to see that this is an interval assignment for G. ▯

Using Theorem 1, we now get the following sufficient conditions on D for its competition graph to be an interval graph.

Corollary 2. Let D be a digraph of a food web. If the vertices of D can be labeled so that for any vertex x, the prey of x are consecutively labeled, then the competition graph of D is an interval graph.

Proof. Label the vertices of D to satisfy the hypothesis of the theorem. Then, if A is the adjacency matrix of D, the rows of A have the consecutive ones property. Hence, by Theorem 1, RG(A) is an interval graph. ▯

This can be restated in terms of a cover of D as follows. An instar in a digraph D is a bipartite subdigraph of D of the form $C_k \leftrightarrow u_k$ where C_k is the set of all vertices that have an arc to vertex u_k. Obviously the set of all instars covers the arcs of D. We will say that a labeling of an instar cover is consecutive if whenever vertex x is in C_i and C_j for $i < j$, then for all m such that $i < m < j$, x is in C_m. Corollary 2 is then equivalent to the following statement.

Corollary 3. If the instar cover of a digraph D has a labeling that is consecutive, then the competition graph of D is an interval graph. ▯

The condition in the above theorem is not necessary as is seen from examining the digraph and its competition graph of Figure 1. Clearly the competition graph is an interval graph, but the set $C = \{\{2,3,4\}, \{3,4,6\}, \{4,5\}, \{5,6\}\}$ does not have a labeling that is consecutive.

To get necessary and sufficient conditions we need another type of cover of D. A set $S = \{C_1,...,C_r\}$ of vertices of D will be called a competition cover of D if

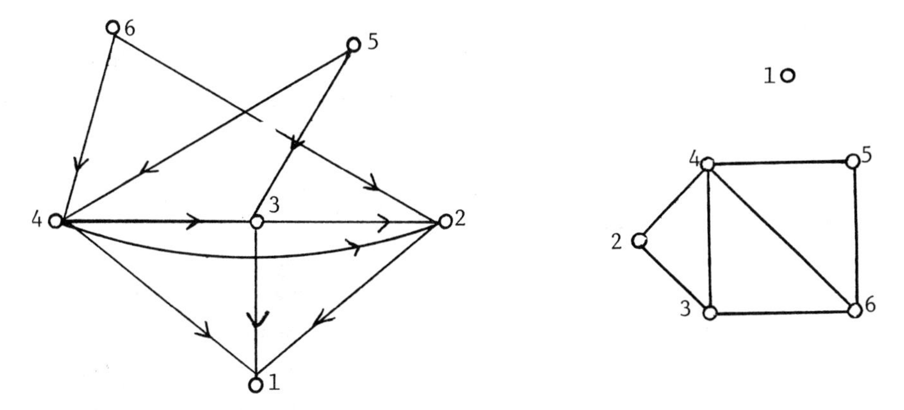

Figure 1. The competition graph G of a digraph D.

the following conditions are satisfied:

1) i, j $\in C_m$ implies that there exists a vertex k such that (i,k) and (j,k) are arcs,
2) if (i,k) and (j,k) are arcs in D, then i, j $\in C_m$ for some m.

Observe that the set of all C_k in an instar cover form a competition cover of D. We now get the following characterization if digraphs whose competition graph is interval.

Theorem 4. The competition graph G of a digraph D is an interval graph if and only if D has a competition cover S which has a ranking that is consecutive.

Proof. The idea of the proof is that the sets C_i in S correspond to cliques in G. Suppose G is an interval graph. The set $S = \{C_1,...,C_r\}$ of maximal cliques of G covers all the edges of G so clearly S is a competition cover of D, and S has a ranking that is consecutive by a result of Fulkerson and Gross [4].

Suppose D has a competition cover $S = \{C_1, \ldots, C_r\}$ which has a ranking that is consecutive. Then each C_i is also a clique in G, and since S is a competition cover of D, the set S covers all the edges in G. We can then use the same method that we used in Theorem 1 to find an interval assignment for G. □

It is important to note that the purpose of the above theorems is to provide insight into the structure of D and A when G = RG(A) is an interval graph. In general, it is more efficient to use standard algorithms on G to determine if it is an interval graph than to search for a competition cover in D.

Next we consider if an acyclic digraph D could have a list of forbidden generated subdigraphs. As observed by Steif [17], this is not possible. The digraph D in Figure 2 is acyclic, and its competition graph, $K_6 \cup I_1$, is interval. However, the generated subdigraph D - {g} has competition graph $Z_4 \cup I_2$, which is not interval.

Another approach to finding forbidden characterizations for acyclic digraphs with interval competition graphs is to use the sink and source food webs introduced by Cohen [2]. A sink induced subdigraph H of a digraph D, is a generated subdigraph with the following additional property: if $x \in H$ and (x,y) is an arc in D, then $y \in H$. A source induced subdigraph H of a digraph D is a generated subdigraph with the following additional property: if $y \in H$ and (x,y) is an arc in D, then $x \in H$. Cohen [2] obtained the following characterization of digraphs with interval competition graphs.

Theorem 5. (Cohen [2]) An acyclic digraph has a competition graph that is interval if and only if every sink induced subdigraph does.

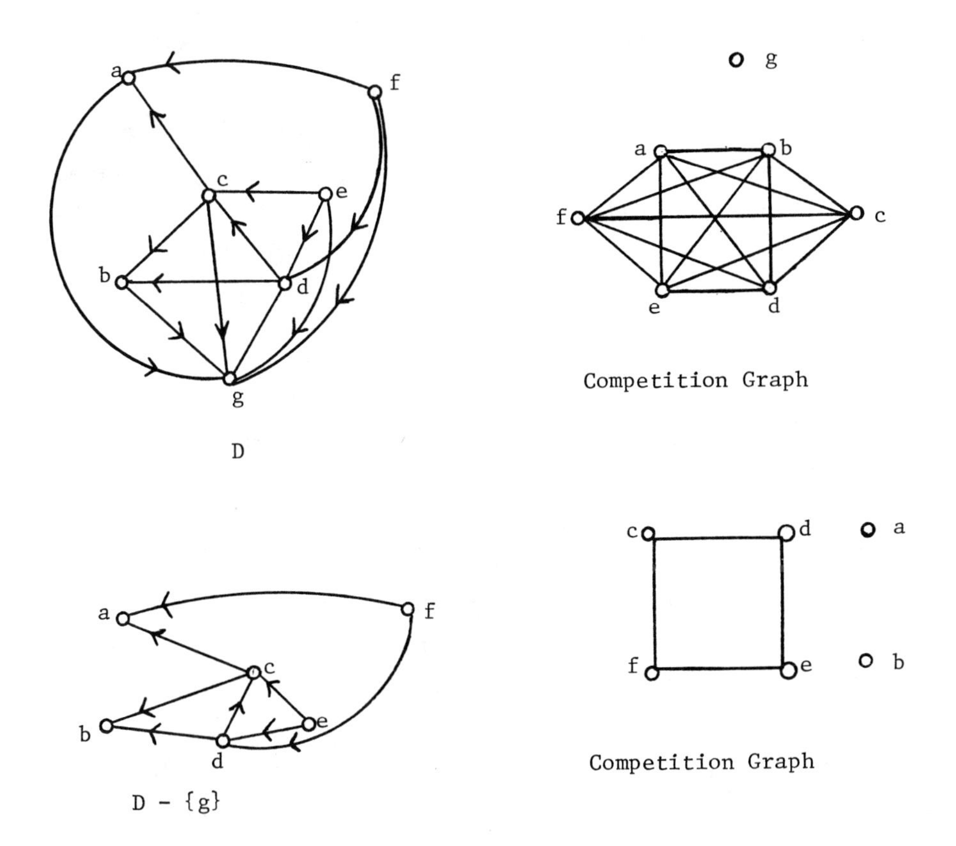

Figure 2. Two digraphs and their competition graphs.

It is possible for a digraph to have an interval competition graph while some source induced subdigraph contained in it has a competition graph which is not an interval graph (see Roberts [9] for an example). However, source induced subdigraphs provide the dual to Theorem 5 for common enemy graphs.

Theorem 6. An acyclic digraph D has a common enemy graph that is interval if and only if every source induced subdigraph does.

Proof. Observe that since D is a source induced subdigraph of itself, the sufficiency is obvious. So suppose D has an interval competition graph G and H is a source induced subdigraph of D with common enemy graph Y. Let y_1 and y_2 be vertices in Y. Then y_1 and y_2 have a common enemy in D if and only if they have a common enemy in H. Hence Y is a generated subgraph of G, so the result follows since every generated subgraph of an interval graph is an interval graph. ▯

From the above discussion we see that for both the competition and common enemy graphs to be interval, the conditions of both Theorems 5 and 6 have to be satisfied. These theorems are useful in showing that a competition or common enemy graph are not interval if one can find a forbidden sink induced or source induced subdigraph. This situation is illustrated in Figure 3. Here H is a sink induced subdigraph of D with competition graph G_1, which is not an interval graph, so G is not an interval graph, and F is a source induced subdigraph of D with common enemy graph E_1 which is not an interval graph, so E is not an interval graph. Observe that G_1 and E_1 are induced subgraphs of G and E respectively.

Steif [17] has pursued these ideas as follows. A forbidden sink induced subdigraph list for property P is a list L of digraphs such that a digraph D has property P if and only if D does not contain any element of L as a sink induced subdigraph. Steif [17] proves the following result.

Theorem 7. (Steif [17]). There exists a forbidden sink induced subdigraph list for (acyclic) digraphs with interval competition graphs.

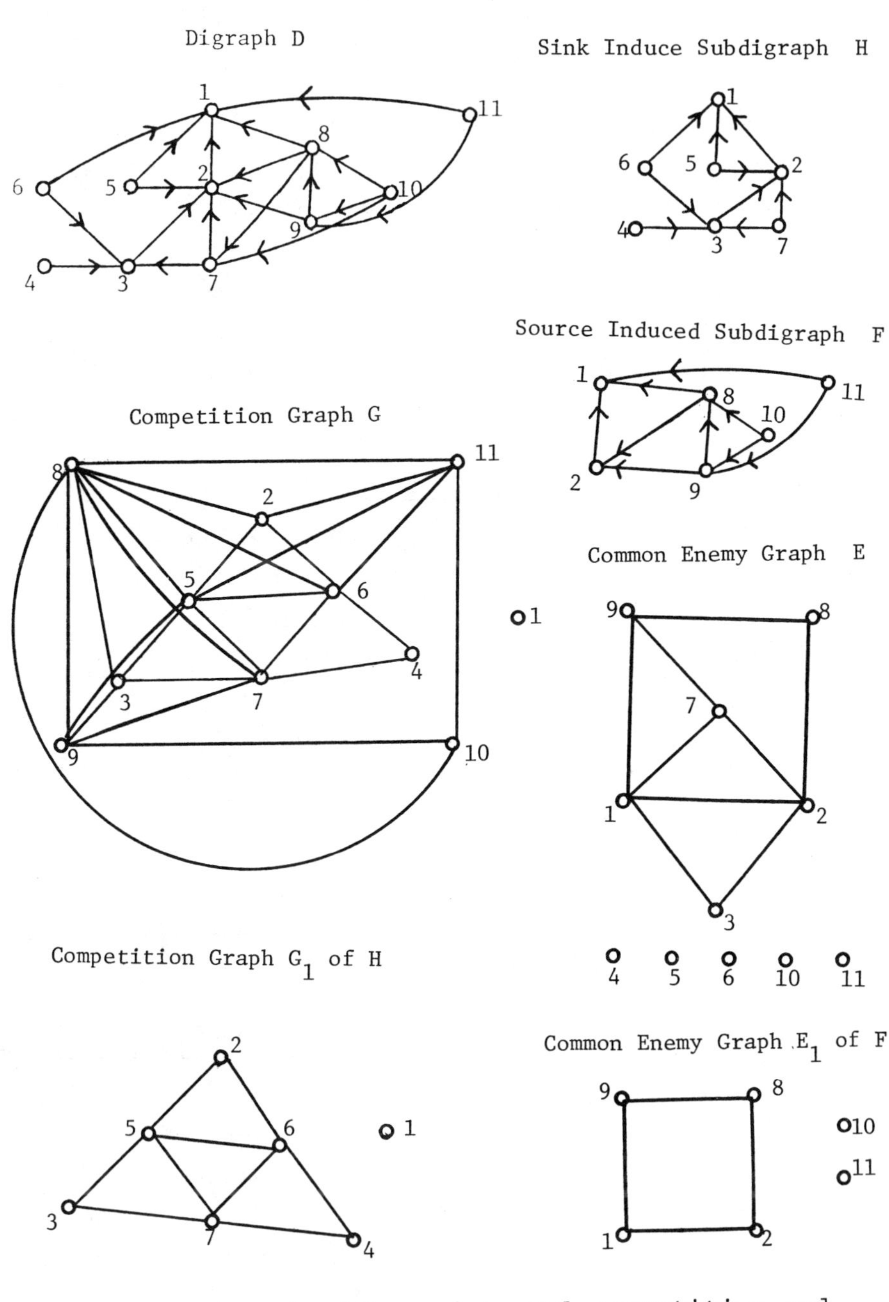

Figure 3. Digraph with noninterval competition and common enemy graphs.

To find such a list appears to be a difficult problem. If D is acyclic, then by Theorem 10.1 of [7], the vertices of D can be labeled so that A is strictly lower triangular. Lundgren and Maybee [9] give a method for constructing all digraphs having a given graph G as their competition graph by using edge covers of G to construct the adjacency matrix of D. We would then have to find digraphs having competition graphs with the cycle Z_n, $n \geq 4$ as a generated subgraph or which contain an asteroidal triple (see Lekkerkerker and Boland [8]). Hopefully, one could reduce this problem to finding a few infinite families. Theorem 4 may also be helpful in trying to find minimal digraphs on this list. For example, digraphs D_1 and D_2 in Figure 4 both have Z_4

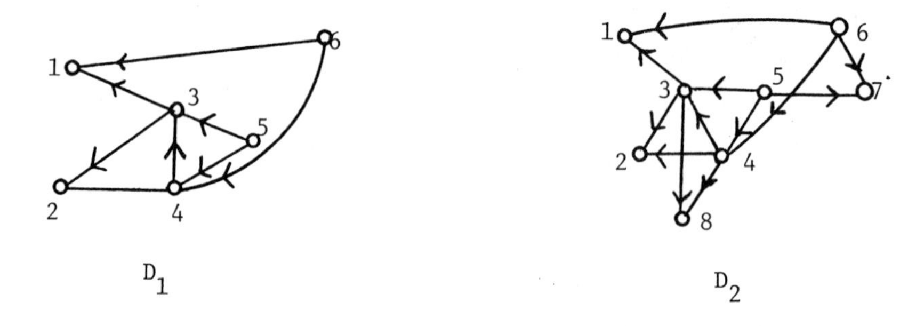

Figure 4. Digraphs with Z_4 as generated subgraphs of their competition graphs.

as a generated subgraph of their competition graph, but D_2 does not require the arcs to vertices 7 and 8 to get Z_4 while D_1 needs all of its arcs. This example also illustrates that perhaps Theorem 4 should be used to find some modification of the notion of sink induced subdigraph since $D_2 - \{7,8\}$ is not a sink induced subdigraph.

Another approach to this problem is to consider a more restricted class of digraphs. Lundgren and Maybee [10] have shown that the upper bound graph, studied by McMorris and Zaslavsky [12] and McMorris and Myers [11], is a competition graph of a digraph that is reflexive, antisymmetric, and transitive. McMorris (personal communication) has suggested that first finding all posets whose upper bound graphs are interval may help to solve the problem.

REFERENCES

1. J. E. Cohen, Interval graphs and food webs: A finding and a problem. Rand Corporation Document 17696-PR, Santa Monica, CA (1968).
2. __________ , Food Webs and Niche Space. Princeton University Press, Princeton (1978).
3. R. D. Dutton and R. C. Brigham, A characterization of competition graphs. Discrete Applied Math. (1982), to appear.
4. D. R. Fulkerson and O. A. Gross, Incidence matrices and interval graphs. Pacific J. Math., 15 (1965) 835-855.
5. J. H. Greenberg, J. R. Lundgren and J. S. Maybee Graph theoretic methods for the qualitative analysis of rectangular matrices. SIAM J. Alg. Disc. Meth. 2, (1981), 227-239.
6. ____________ , Inverting graphs of rectangular matrices. (1981), to appear.
7. F. Harary, R. Z. Norman and D. Cartwright, Structural Models: An Introduction to the Theory of Directed Graphs. Wiley, New York (1965).

8. C. B. Lekkerkerker and J. C. Boland, Representation of a finite graph by a set of intervals on the real line. Fund. Math.51 (1962), 45-64.
9. J. R. Lundgren and J. S. Maybee, A characterization of graphs of competition number m. Discrete Applied Math. (1982), to appear.
10. _________, A characterization of upper bound graphs. (1982), to appear.
11. F. R. McMorris and G. T. Myers, Some uniqueness results for upper bound graphs. Discrete Math. (1982), to appear.
12. F. R. McMorris and T. Zaslavsky, Bound graphs of a partially ordered set. J. Combin. Inform. System Sci. (1982), to appear.
13. R. J. Opsut, On the computation of the competition number of a graph. SIAM J. Alg. Disc. Meth. (1982), to appear.
14. F. S. Roberts, Food webs competition graphs, and the boxicity of ecological phase space. Springer Lecture Notes Math. 642 (1978), 477-490.
15. _________, Graph Theory and Its Applications to Problems of Society. Society for Industrial and Applied Mathematics, Philadelphia, PA (1978).
16. J. S. Roberts and J. E. Steif, A characterization of competition graphs of arbitrary digraphs. Discrete Applied Math. (1982), to appear.
17. J. E. Steif, Frame dimension, generalized competition graphs and forbidden sublist characterizations. (1982), to appear.

EXTREMELY GREEDY COLORING ALGORITHMS

Bennet Manvel

Colorado State University
Fort Collins, CO 80523

ABSTRACT. Graph coloring has an extensive history and theory, but coloring algorithms are a rather recent development. We briefly summarize what is known about such algorithms. Two recent applications which have increased interest in coloring large graphs are presented. Finally, several new algorithms developed by Johri and Matula specifically to color such graphs are presented.

1. INTRODUCTION

A coloring of a graph G can be considered as a function from the set of n vertices V to a set S of "colors" such that adjacent vertices have distinct images. The minimum cardinality of S is the chromatic number $\chi(G)$. Equivalently, it is a partitioning of V into independent sets. The immense amount of attention which graph coloring has received began as efforts toward solution of the recently settled Four Color Problem. It has continued because of the intrinsic interest of the problem, and applicability of the partitioning results obtained. We are concerned here with the efforts, only recently begun, to find efficient ways to actually assign colors to vertices of a graph so that it is colored in $\chi(G)$ colors, or perhaps a few more.

We begin with a brief exposition of two recent applications of coloring as a partitioning method. A summary of existing coloring algorithms, emphasizing methods for fast approximate colorings, follows. None of the algorithms found in the literature do a good job of coloring large ($n \geq 1000$) graphs, and we try to explain that failure. We then present several brand-new algorithms written by Johri and Matula [15] which exploit properties of large graphs to do a better coloring job, and close with several open questions.

2. APPLICATIONS OF COLORING

Graph coloring, of course, originated as an applied problem: coloring the faces of a map. Unfortunately, there is no evidence that cartographers were at all interested in the results obtained! More recently, colorings have been used in solving certain problems of scheduling (Peck and Williams [20], Brown [3]), routing (Tucker and Bodin [23]), and others [1], [8]. Although there are clearly fewer applications of colorings than there are of flow or path algorithms, two recent articles show that coloring methods can be applied in unexpected ways.

In 1976 Garey, Johnson and So of Bell Labs published an application of graph coloring to circuit testing [11]. When a circuit board is constructed, connections are made on the reverse side of the board joining certain nodes into nets which are electrically common, see Figure 1. There is a positive probability

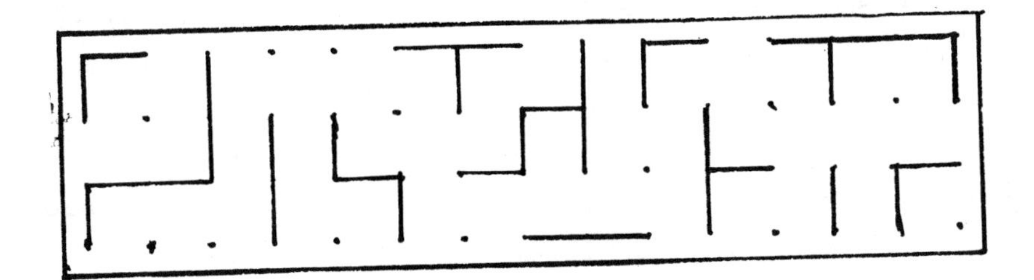

Figure 1

that some extraneous connections are made by solder carelessly applied. An obvious way to test for such shorts is to test every pair of nets. Clearly, however, pairs of nets which are some distance apart on the board need not be tested. Thus it is reasonable to make the nets the vertices of a graph, and define adjacency in some way so that vertices are very unlikely to be shorted together unless they are adjacent in the graph (that is, close on the circuit board). Then one test for every edge of that graph will almost certainly uncover all shorts. In fact, however, a much more efficient testing scheme is possible.

To test efficiently, first color the graph in k colors (Garey, Johnson and So show that 12, 8 or 5 colors will suffice, depending on how stringent the adjacency criterion is for the graph). Then construct for each color class a yoke of wires which contact simultaneously all nets in that color class. Testing these external yokes in pairs against each other will then reveal any shorts between vertices which might be adjacent, because they will be in different color classes. By the definition of adjacency, vertices in a single color class cannot short to each other, so only $\binom{k}{2}$ tests will reveal any shorts present.

Notice that this application requires a good deal of work for each type of circuit board: the graph must be constructed and colored and yokes must be made to do the actual testing. This work is well rewarded, however, by the reduction in testing time if many circuit boards of a given type are to be constructed and tested.

Another recent and rather surprising application of coloring methods is contained in work of Curtis, Powell and Reid [6] and Coleman and Moré [5]. If F is a function from R^n to R^m, and F' is the Jacobian of F, it is possible to approximate F' using the equation

$$F'(x)d = F(x+d) - F(x) + o(\|d\|). \tag{1}$$

This can be used to find the entries of F', which is usually a fairly sparse matrix, by careful choice of certain vectors d.

To that end, we define a graph G whose vertices are the columns of F', and say that two vertices are adjacent if and only if the corresponding columns both have an unknown entry in the same row. (The location of non-zero entries in F' is known in advance.) Then a coloring of G specifies a set of vectors d_i, which are the characteristic vectors of the color classes. Those vectors can be used in equation (1) to solve for the entries of F'. For example, if equation (1) looks like

$$F'd_1 = \underbrace{\begin{bmatrix} 0 & ? & 0 & 0 & 0 & ? & \cdots \\ 0 & 0 & ? & 0 & ? & ? & \ddots \\ 0 & ? & 0 & ? & 0 & 0 & \\ ? & 0 & 0 & ? & 0 & 0 & \\ 0 & 0 & ? & ? & 0 & ? & \\ 0 & ? & 0 & ? & 0 & 0 & \cdots \\ ? & 0 & 0 & 0 & 0 & 0 & \cdots \\ 0 & 0 & ? & 0 & ? & 0 & \cdots \\ 0 & 0 & ? & 0 & ? & ? & \cdots \\ ? & 0 & 0 & 0 & ? & 0 & \cdots \\ 0 & ? & 0 & ? & 0 & 0 & \cdots \end{bmatrix}}_{F'} \underbrace{\begin{bmatrix} 1 \\ 1 \\ 1 \\ 0 \\ 0 \\ 0 \\ \cdot \\ \cdot \\ \cdot \\ \cdot \\ \cdot \end{bmatrix}}_{d_1} = (4.71, 2\ 96, -6.28, 11.91, \ldots) + o(3)$$

for $d_1 = (1, 1, 1, 0, 0, 0, \ldots)$ which indicates that v_1, v_2, v_3 are in color class 1, then clearly all unknown entries in columns 1, 2 and 3 can be found. Notice that no two of v_1, v_2, v_3 are adjacent, since there is at most one unknown entry in each row of the first three columns of F'. The number of times that equation (1) needs to be evaluated is just the number of color classes we use for G.

2. EXISTING ALGORITHMS

Many algorithms exist for graph coloring, but they represent only a few different approaches to the problem. Because coloring is an NP-complete problem [16], it is not surprising that no good, fast algorithm is known. (For an excellent exposition of the theory of NP-completeness, see [10].) Algorithms for finding χ exactly, which involve a prohibitive amount of computer time for large problems, are in Hammer and Rudeanu [13], Christofides [4], Pershin [21], Wang [24], and Brelaz [2]. Because we are interested in coloring large graphs, we will concentrate our attention on approximate solutions, based on various heuristics.

Obviously the vertices of a graph can be colored, one by one, as they are encountered, only being careful not to color any vertex the same color as an adjacent one. Such a GREEDY algorithm gives a benchmark against which other heuristics can be measured. All other existing fast coloring algorithms are driven by one or more of the following heuristic assumptions:

A. a vertex of high degree is harder to color than a vertex of low degree,

B. vertices adjacent to the same set of vertices should be colored alike,

C. coloring many vertices with the same color is good.

The simplest (and earliest) such algorithms (besides GREEDY) all rely on A to color the vertices sequentially, i.e., in an order determined before actual coloring begins. Gus Simmons' colorful paper in this collection studies this sequential coloring number. In each case, the next vertex is colored with the smallest possible color. The LARGEST FIRST [25] algorithm orders vertices by descending degree, and then colors greedily. SMALLEST LAST [19] selects a vertex of lowest degree to color last, removes it, and repeats the process. When it finishes pre-ordering, it proceeds to color the vertices, beginning with the last vertex removed, greedily. A method of Williams [26] modifies LARGEST FIRST by considering neighborhood degree sums, and so on, instead of degrees. Dunstan [7] suggests several similar modifications.

All of these degree-based algorithms are sequential, in the sense that they decide, before coloring the first vertex, the order in which the vertices will be colored. They have been modified in various ways to act recursively, as the coloring proceeds (see [15], [17], and [22]). The existing algorithms based on heuristics B and C are fundamentally dynammic, rather than sequential. Wood [27] suggests defining a similarity measure on pairs of vertices, and then examines vertices in pairs in order of decreasing similarity, sometimes coloring and sometimes not coloring, depending on degrees of the pair and whether or not one of the two is already colored. Johnson [14] suggests following heuristic C by selecting, recursively, vertices of low degree to construct large independent sets.

Arguably the best fast coloring algorithm known for coloring reasonably small graphs is the DSATUR (degree of saturation) method, due to Brelaz [2]. It is driven by heuristics A and B, as follows. The color-degree of v is the number of colors used so far on vertices adjacent to v. DSATUR repeatedly chooses a vertex of largest color-degree, and assigns to it the smallest possible color. If two vertices have equal color-degrees, the one adjacent to most uncolored vertices is colored first.

All of these algorithms do a reasonable job of coloring small graphs. DSATUR, in fact, colors even 100-vertex graphs using only a few extra colors. Many applications of coloring (in particular, the two we explained in Section 1) sometimes require coloring larger graphs, with perhaps 1000 vertices or more. Of course, these are exactly the type of graph for which exact coloring algorithms are totally impractical. It has been noted [18] that none of the heuristic algorithms, including DSATUR, does a very good job on such large graphs.

This failure in coloring large graphs may be due, at least partially, to the way in which χ is determined in large and small graphs. Two different lower bounds drive χ. Clearly, $\chi(G)$ is at least as large as the size $\omega(G)$ of the largest complete subgraph of G. For small graphs $\chi(G)$ and $\omega(G)$ in fact tend to be fairly close in value. On the other hand, if $\beta(G)$ is the size of the largest independent subgraph of G, then clearly $\chi(G)$ is at least $n/\beta(G)$. For small graphs $n/\beta(G)$ is usually much smaller than $\chi(G)$, but for large graphs, the $n/\beta(G)$ lower bound tends to be far closer to $\chi(G)$ than $\omega(G)$ is.

Matula has calculated the expected value of $\beta(G)$ for random graphs G with edge-probability $p = .5$. It turns out that $\beta(G)$ can be predicted with near-certainty. For example, for G with 1000 vertices and $p = .5$, the probability that $\beta(G) = 15$ (no more, no less) is more than 80%. With such information for graphs of all sizes up to 1000, a likely estimate for χ can be found. Beginning with G on 1000 vertices, assume an independent set of size 15 can be found, and delete it. The table in [15] says that an 885 vertex graphs is also likely to have an independent set of size 15. Deleting that set and continuing in the same way, we obtain an estimate $\tilde{\chi}(G)$, for the chromatic number of G. Table 1 gives expected values for $\omega(G)$, $n/\beta(G)$ and $\tilde{\chi}(G)$ for graphs on various number of vertices. Notice how $\omega(G)$ is close to $\tilde{\chi}$ for small n, but $n/\beta(G)$ becomes a better bound for n large. This suggests a difference in character of small and large graphs which was first pointed out in [18].

TABLE 1

Edge Probability = .5

n	$\omega(G)$ $(=\beta(G))$	$n/\beta(G)$	$\tilde{\chi}$
6	3	2	3
15	5	3	5
42	7	6	10
130	10	13	19
312	12	26	36
1000	15	67	85

Direct evidence that a difference of character, and not mere size, is causing trouble for existing heuristics is in a paper of Leighton [17]. He generates large graphs in a psuedo-random way which assures that $\chi(G)$ in fact equals $\omega(G)$. On those graphs his recursive version of the largest-first algorithm performs rather well.

When tested on large graphs generated in a pseudo-random way which assured that $\chi(F)$ was $n/\beta(G)$, none of the published algorithms performed that impressively [18].

3. NEW ALGORITHMS

Johri and Matula have responded to the problems of coloring large graphs with a set of new algorithms, oriented toward finding large independent sets [15]. Their methods are less systematic than previous algorithms, utilizing the capabilities of computers for exhaustive searches and probabalistic results in ingenious ways.

The basic method for all these "greedy-exhaustive" algorithms is to look for the largest independent set which can be expected (given the size and edge-density of the graph), and then delete that set from the graph. For small graphs such a search can be carried out exhaustively. For graphs with more than about 100 vertices exhaustive search is impractical, so a two-phase process is used. The first phase selects an independent set I_1 of vertices large enough to assure that the remaining set R of vertices not adjacent to any member of I_1 is smaller than some preset threshhold T. The second phase then searches exhaustively in R for a largest expected independent set, I_2. (In order to assure that the edge density of the graph does not increase too much when $I_1 \cup I_2$ is deleted, I_2 is selected to cover as many edges as possible.) $I_1 \cup I_2$ is then used as the next color class. Clearly this general method can be fine-tuned in various ways, and thus there is a hierarchy of algorithms, progressively more complex, slower, and utilizing fewer colors.

The simplest such algorithm, GE1, uses random selection of vertices to find I_1 in phase 1. In GE2, a sampling procedure is employed to increase the size of the independent set $I_1 \cup I_2$ found at each stage. A desired independent set size is calculated, and if $I_1 \cup I_2$ is not that large, that set is rejected and both phases repeated to find a new I_1 and I_2. After a certain number of failures, the desired independent set size is reduced and the process repeated. Finally, in the most complex form of this algorithm, GE3, the random selection of I_1 is refined to a sampling procedure, which adds vertices to I_1 by examining several candidates and selecting the one with largest average degree on its adjacent vertices. Also, GE3 abandons the method of GE1 when fewer than 80 vertices remain to be colored, and searches exhaustively for independent sets of largest expected size.

In the particular implementations of these algorithms for which data is given in Table 2 (extracted from Table 12 of Johri and Matula), T was set at 44. The value T = 64 was also tried for GE1, but it was found that the sampling procedure of GE2 was a more efficient way to improve coloring performance. All of the data given is average performance on a set of 10 random graphs on 1000 vertices, with edge probability p = .5. Time is CPU seconds on a CDC 6600.

TABLE 2

Algorithm	Average Colors Used	Average Time
RANDOM	127.3	100.14
LARGEST FIRST	122.7	101.31
SMALLEST LAST	124.3	126.73
DSATUR	115.8	111.36
GE1	105.2	432.81
GE2	100.1	1128.24
GE3	95.9	3212.17

Since $\tilde{\chi}$, the likely value for χ, is 85 for such graphs, the greedy exhaustive algorithms are indeed effective coloring methods. The gap in quality and time between DSATUR and GE1 can be filled to some degree by recursive versions of the simpler algorithms [15], but the gain seems small for the extra time expended. The timing data is, of course, very dependent on implementation. In particular, data in [18] indicates that DSATUR is only half as fast as LARGEST FIRST on 1000 vertex, $p = .5$ graphs, which is not what Table 2 indicates at all. Clearly, however, the greedy exhaustive algorithms color large graphs quite well.

4. OPEN QUESTIONS

The following two questions are fundamental to the theory of coloring algorithms.

1. Does there exist a polynomial time algorithm which colors all graphs G in $k\chi(G)$ colors, for some k?
2. Does there exist a polynomial time algorithm which colors large graphs well, on average?

Garey and Johnson have shown that coloring in $k\chi(G)$ colors for $k < 2$ in NP-complete [9], and conjecture that is true for every k. Thus question 1 is surprisingly difficult. For question 2, it is known that RANDOM uses at most $2\chi(G)$ colors, almost always [12], but probabilistic results are difficult for more complex algorithms. Clearly some of the algorithms seem to be better than RANDOM, but how could that be proved.

The other questions relate directly to the new algorithms we have described.

3. Are the GE methods as efficient for sparse graphs with edge-probability $p = .5$?

Large graphs in applications, such as the ones we discussed in Section 2, are usually sparse. In fact, $p = .01$ might be an appropriate choice, rather than $p = .5$. They also frequently have special structure, for example, the planarity of a subdivision used in a finite element problem. Algorithms which recognize the special character of those graphs are needed.

4. Are the probabilistic and sampling methods employed so effectively in GE1, 2 and 3 applicable to other approximate graph algorithms?

For example, could graph isomorphism be tested by random selection of subgraphs of G and search for corresponding subgraphs in H? Several other problems might be even more natural for such an approach.

REFERENCES

1. F.F. Atstonas, K.I. Plukas, A method of minimization of microprograms for electronic digital computers. Avtomat. i Vycisl. Tehn. 4(1971) 10-16.
2. D. Brelaz, New Methods to color the vertices of a graph. Comm. ACM 22(1979) 251-256.
3. J. R. Brown, Chromatic scheduling and the chromatic number problem. Management Science 19(1972) 456-463.
4. N. Christofides, An algorithm for the chromatic number of a graph. Comp. J. 14(1971) 38-39.

5. T.F. Coleman and J.J. Moré, Estimation of sparse jacobian matrices and graph coloring problems. ANL-81-39, Argonne Nat'l. Lab., (June 1981).
6. A.R. Curtis, M.J.D. Powell and J.K. Reid, On the estimation of sparse jacobian matrices. J. Inst. Maths. Appl. 13(1974) 117-119.
7. F.D.J. Dunstan, Sequential colorings of graphs. Cong. Num. 15(1976) 151-158.
8. G.F. Fricnovic, Coding of the internal states of asynchronous finite automata by a p-code of minimal length. Theory of Finite Automata and its Applications 1(1973) 22-34, 64-65.
9. M.R. Garey and D.S. Johnson, The complexity of near optimal graph coloring. J.A.C.M. 23(1976) 43-49.
10. M.R. Garey and D.S. Johnson, Computers and Intractability. Freeman, San Francisco (1979).
11. M.R. Garey, D.S. Johnson and H.C. So, An application of graph coloring to printed circuit testing. IEEE Trans. on Circuits and Systems 23 (1976).
12. G.R. Grimmett and C.J.H. McDiarmid, On colouring random graphs. Math. Proc. Comb. Phil. Soc. 77(1975) 313-324.
13. P.L. Hammer and S. Rudeanu, Boolean Methods in Operations Research. Springer-Verlag, New York (1968).
14. D.S. Johnson, Worst case behavior of graph coloring algorithms. Cong. Num. 19(1974) 513-527.
15. A. Johri and D.W. Matula, Probabilistic bounds and heuristic algorithms for coloring large random graphs, to appear.
16. R.M. Karp, Reducibility among combinatorial problems. Complexity of Computer Computations (R.E. Miller and J.W. Thatcher, eds.), Plenum Press, New York (1972) 85-104.

17. F.T. Leighton, A graph coloring algorithm for large scheduling problems. J. of Research. National Bureau of Standards 84(1979) 489-496.

18. B. Manvel, Coloring large graphs. Proc. of the 1981 S.E. Conf. on Graph Theory, Combin. and Computer Science, to appear.

19. D.W. Matula, G. Marble and J.D. Isaacson, Graph coloring algorithms. Graph Theory and Computing (R.C. Read, ed.), Academic Press, New York (1972) 109-122.

20. J. Peck and M. Williams, Examination scheduling. Comm. ACM 9 (1966) 433-34.

21. O. Persin, An algorithm for determining the minimum coloring of a finite graph. Engineering Cybernetics 11(1973) 980-985.

22. A. Tehrani, Un algorithme de coloration. Cahiers Centre Etudes Recherche Oper. 17(1975) 395-398.

23. A.C. Tucker and L. Bodin, A Model for municipal street sweeping operations. Case Studies of Applied Mathematics, Math. Assoc. of America, Washington (1976) 251-295.

24. C.C. Wang, An algorithm for the chromatic number of a graph. J.A.C.M. 21 (1974) 385-391.

25. D.J.A. Welsh and M.B. Powell, An upper bound to the chromatic number of a graph and its application to time-tabling problems. Comp. J. 10(1967) 85-86.

26. M.R. Williams, The coloring of very large graphs. Combinatorial structures and their Applications, Gordon and Breach, New York (1970) 477-478.

27. D.C. Wood, A technique for coloring a graph applicable to large-scale time-tabling problems. Comp. J. 12(1969) 317-319.

PANCYCLIC AND BIPANCYCLIC GRAPHS - A SURVEY

John Mitchem and Edward Schmeichel

San Jose State University
San Jose, CA 95192

ABSTRACT. A n-vertex graph G is called pancyclic if it contains cycles of all lengths k, for $3 \leq k \leq n$. We survey conditions which imply that G is pancyclic, and consider how these conditions might be weakened if G is assumed hamiltonian. Analogous questions for bipartite graphs are also considered.

1. PANCYCLIC GRAPHS

In this survey, we consider only simple, undirected graphs. Our terminology and notation will be standard except as indicated. A good reference for any undefined terms is [5].

A graph G on n vertices is called pancyclic if it contains cycles of every length k, for $3 \leq k \leq n$. The starting point of the questions we consider is the following, loosely worded "metaconjecture" of Bondy [2]:

BONDY'S METACONJECTURE. Almost any nontrivial condition on a graph which implies that the graph is hamiltonian also implies that the graph is pancyclic. (There may be a simply described family of exceptional graphs.)

Various sufficient conditions for a graph to be hamiltonian have been given, but among the easiest and most appealing are the following.

Theorem A. (Ore [9]). Let G be a graph on $n \geq 3$ vertices. If $q > \binom{n-1}{2} + 1$, then G is hamiltonian.

Theorem B. (Dirac [3]). Let G be a graph on $n \geq 3$ vertices. If $\delta(G) \geq n/2$, then G is hamiltonian.

One sees that the bounds in these theorems cannot be improved in general by considering $(K_{n-2} \cup K_1) + K_1$ for Theorem A and $K(\lfloor (n-1)/2 \rfloor, \lceil (n+1)/2 \rceil)$ for Theorem B.

In line with Bondy's Metaconjecture, the following results are known.

Theorem C. (Bondy [1]). Let G be a graph on $n \geq 3$ vertices. If $q > \binom{n-1}{2} + 1$, then G is pancyclic.

Theorem D. (Bondy [1]). Let G be a graph on $n \geq 3$ vertices. If $\delta(G) \geq n/2$, then G is pancyclic or $K(n/2, n/2)$.

It appears that most of the strength of the conditions in Theorems C and D is necessary to guarantee the existence of a hamiltonian cycle in G. In fact, this is the underlying idea of Bondy's Metaconjecture. It seems an interesting question therefore how much the bounds in Theorems C and D might be lowered if one makes the additional assumption that G is hamiltonian.

For Theorem C, a satisfying answer to the above question has been found. The following result was first conjectured by Erdös.

Theorem E. (Häggkvist, Faudree, and Schelp [4]). Let G be a hamiltonian graph on n vertices. If $q > \lfloor (n-1)^2/4 \rfloor + 1$, then G is pancyclic or bipartite, and the bound is best possible.

We prove only that the bound in Theorem E is best possible when n is odd. Consider $K((n-1)/2, (n-1)/2)$, with distinguished adjacent vertices x, y. Remove edge xy, and create a new vertex z adjacent to precisely x and y. The resulting graph G has n vertices and $\lfloor (n-1)^2/4 \rfloor + 1$ edges. Moreover G is hamiltonian but neither bipartite (G has a 5-cycle) nor pancyclic (G contains no 3-cycle). Slightly more complicated examples can be constructed to show that the bound is also best possible in the case when n is even. (Unfortunately, the example in [4] is not correct in this case.)

Lowering the bound in Theorem D when G is assumed hamiltonian seems much more difficult. To our knowledge, the n/2 bound in Theorem D has never been lowered at all! We have only the following conjecture to offer.

Conjecture 1. Let G be a hamiltonian graph on n vertices. If $\delta(G) \geq (2n+1)/5$, then G is pancyclic or bipartite, and the bound is best possible.

We show that the bound in Conjecture 1 cannot be improved when $n \equiv 0 \pmod 5$. Take $r = n/5$, and let G be the wreath product of $\bar{K}_r$ and C_5 (see Fig. 1). It is easily verified that G is 2/5 n-regular and hamiltonian, but neither bipartite (G contains a 5-cycle) nor pancyclic (G contains no 3-cycle).

As a partial result towards Conjecture 1, Erdös and Locke (private communications) have proven the following

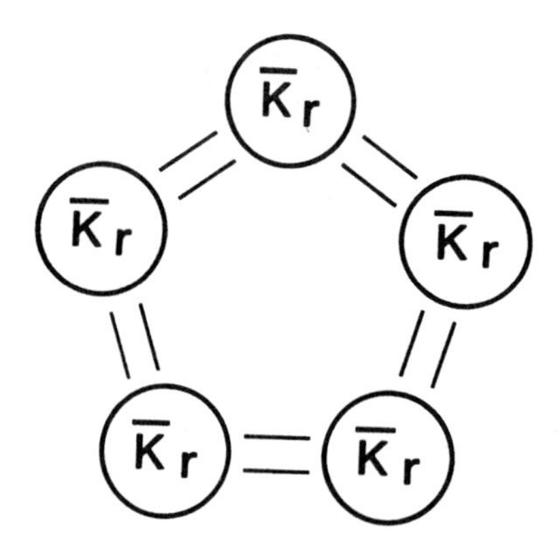

Figure 1.

result: If G is a graph on $n \geq 3$ vertices and $\delta(G) \geq (2n+1)/5$, then G contains a k-cycle for $3 \leq k \leq \lfloor (n+18)/10 \rfloor$: (There is no need to assume G is hamiltonian).

3. BIPANCYCLIC GRAPHS

A bipartite graph on 2n vertices is called bipancyclic if it contains cycles of every even length k, for $4 \leq k \leq 2n$. One could easily formulate a statement for bipartite graphs analogous to Bondy's Metaconjecture for general graphs.

We begin by stating two conditions for a bipartite graph to be hamiltonian analogous to Theorems A and B.

Theorem F. (Moon-Moser [8]). Let G be a balanced bipartite graph on 2n vertices. If $q > n(n-1) + 1$, then G is hamiltonian, and the bound is best possible.

Theorem G. (Moon-Moser [8]). Let G be a balanced

bipartite graph on $2n \geq 4$ vertices. If $\delta(G) \geq (n+1)/2$ then G is hamiltonian, and the bound is best possible.

To prove that the bound in Theorem F is best possible, take $K(n-1,n)$ and create a new vertex adjacent to exactly one vertex in the bipartition set with n vertices. The bound in Theorem G is seen to be best possible by considering $2K(n/2,n/2)$.

In line with our metaconjecture for bipartite graphs, we have the following results.

Theorem 1. Let G be a balanced bipartite graph on 2n vertices. If $q > n(n-1) + 1$, then G is bipancyclic.

Proof. The theorem is certainly true for $n = 2$, and so we proceed by induction on n.

Let G have bipartition $X \cup Y$. Choose a vertex x in X which is not adjacent to at least one vertex in Y. Since $|E(G)| > n(n-1) + 1$, x must be adjacent to at least one vertex $y \in Y$.

Consider $G' = G-x-y$. Note that G' is a balanced bipartite graph on $2(n-1)$ vertices, and that

$$|E(G')| > n(n-1)+1 - 2(n-1) = (n-1)(n-2)+1.$$

By the induction hypothesis, G' is bipancyclic. So to show that G is bipancyclic, it suffices to show that G is hamiltonian. But this is the result of Theorem F. ▯

Theorem H. (Mitchem-Schmeichel [6]). Let G be a balanced bipartite graph on $2n \geq 4$ vertices. If $\delta(G) \geq (n+1)/2$, then G is bipancyclic.

We next consider the question of how much the bounds in Theorems 1 and H might be lowered if we make the additional assumption that G is hamiltonian.

We have the following result which reduces the bound in Theorem 1 when G is hamiltonian.

Theorem I. (Mitchem-Schmeichel [7]). Let G be a hamiltonian bipartite graph on 2n vertices. If $q > n^2/2$, then G is bipancyclic.

The authors, however, feel that the bound given in Theorem I is probably not best possible. Indeed we make the following

Conjecture 2. Let G be a hamiltonian bipartite graph on 2n vertices. If $q \geq (1+n/2)^2$, then G is bipancyclic and the bound is best possible.

To see that the bound in Conjecture 2 would be best possible, consider the graphs shown in Fig. 2. It is

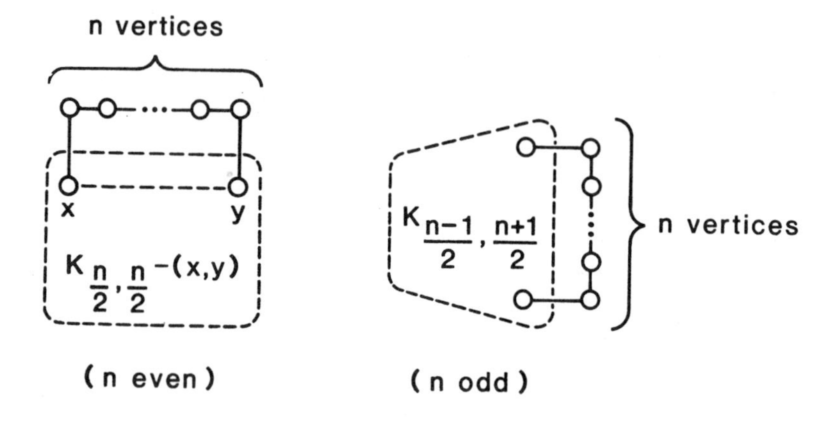

Figure 2.

easily verified that the graphs are hamiltonian with $\lfloor(1+n/2)^2 - 1/4\rfloor$ edges, and that the only missing even cycle length is (n+2) when n is even or (n+1) when n is odd.

Reducing the bound in Theorem H when G is

hamiltonian seems quite difficult. As was the case with Theorem D, we know of no result which lowers the bound in Theorem H by any amount even if G is assumed hamiltonian. We propose only the following statement.

Conjecture 3. Let G be a hamiltonian bipartite graph on 2n vertices. If $\delta(G) > (1 + \sqrt{4n-3})/2$, then G is bipancyclic and the bound is best possible.

We have shown [10] that the bound in Conjecture 3 certainly cannot be improved when $n = p^2 + p + 1$, where p is a prime. For each such n, an extremal graph turns out to be the point-line incidence graph of a certain projective plane of order p. Each such graph is $(1 + \sqrt{4n-3})/2)/2$ - regular, hamiltonian and contains a cycle of every even length k for $6 \leq k \leq 2n$; i.e., the graph is mssing only a 4-cycle.

NOTE ADDED IN PROOF: Conjecture 1 has recently been settled in the affirmative by four graph theorists in Paris (D. Amar, E. Flandrin, I. Fournier, and R. Germa, Pancyclism in hamiltonian graphs (to appear)).

REFERENCES

1. J.A. Bondy, Pancyclic graphs I. J. Combin. Theory 11 (1971), 80-84.
2. J.A. Bondy, Pancyclic graphs: Recent results. Infinite and Finite Sets (1973), 181-187.
3. G.A. Dirac, Some theorems on abstract graphs. Proc. London Math. Soc. 2 (1952), 69-81.

4. R. Haggkvist, R.J. Faudree and R.H. Schelp, Pancyclic graphs-connected ramsey number. Ars Combin. 11 (1981), 37-49.

5. F. Harary, Graph Theory. Addison-Wesley, Rading (969).

6. J. Mitchem and E. Schmeichel, Bipartite graphs with cycles of all even lengths. J. Graph Theory 6 (1982), 429-439.

7. J. Mitchem and E. Schmeichel, Edge conditions and cycle structure in bipartite graphs (to appear).

8. J.W. Moon and L. Moser, On hamiltonian bipartite graphs. Israel J. Math. 1 (1963), 163-165.

9. O. Ore, Arc coverings of graphs. Ann. Mat. Pura Appl. 55 (1961), 315-321.

10. E. Schmeichel and J. Mitchem, On the cycle structure of finite projective planes (to appear).

HOW TO MINIMIZE THE LARGEST SHADOW OF A FINITE SET

Allen J. Schwenk
U. S. Naval Academy
Annapolis, MD 21402

ABSTRACT. We consider projections of a finite set of n points in Euclidean d-space R^d onto the various s-dimensional hyperplanes R^s obtained by discarding $d-s$ coordinates and keeping the other s. It has been shown previously that the geometric mean of these $\binom{d}{s}$ projections is at least $n^{s/d}$. In this article we examine the problem of finding an optimal n-set whose maximum projection size is as small as possible. We solve the problem for every n when $d = 3$ and $s = 2$. An important lemma is the demonstration that, for $k^3 \leq n < (k+1)^3$, among the optimal n-sets there is at least one that contains an order k cube as a subset and is itself a subset of an order $k + 1$ cube. The minimum projection number attains the lower bound of $n^{2/3}$ whenever $n = k^3$ but for large n it drifts to $n^{2/3} + 3/8\ n^{1/3}$ infinitely often.

The problem in higher dimensions seems to be much more difficult. For arbitrary d and $s = d - 1$ we construct a set for certain very special values of n which we believe to be optimal. We do not even have a good candidate for an optimal set for arbitrary n.

1. THE GENERAL DIMENSIONAL PROBLEM

In [3] we derived a bound on the geometric mean of the size of certain projections of a finite set in d-dimensions. More precisely, given any n points in Euclidean d-space R^d and the $\binom{d}{s}$ possible projections onto s-dimensional hyperplanes R^s defined by selecting s of the coordinate axes, we found that the geometric mean of the cardinalities of these $\binom{d}{s}$ projections is at least $n^{s/d}$. Moreover, this bound is attained for every choice of n, s, and d.

This problem actually arose [1] in the analysis of searching a computer file containing n records each comprised of d items or "keys". By specifying s of these keys the records are partitioned into a number of equivalence classes. That is, two records are equivalent if they have identical keys among the s selected keys. Suppose that we wish to select a subset of s keys which will provide the maximum amount of discrimination among the n records. The mentioned bound guarantees that the geometric mean among all selections of the number of equivalence classes is at least $n^{s/d}$. Thus, some selection exists that does at least this well. For some lists of n records, every selection of s keys might produce very nearly average performance so that no selection exceeds the least integer $\lceil n^{s/d} \rceil$ not less than the bound. Such a collection provides the greatest difficulty if we wish to locate records using a selection of s keys to identify the records since no selection yields more classes than the minimal number guaranteed by the bound. The general and extremely difficult combinatorial problem is to find for each n, s, and d, the most troublesome list of n records in the sense that even the best selection of s keys for that list produces fewer equivalence classes than can be obtained for any other list.

Let us return to the geometric view of the problem. Set X consists of n points in R^d. Let $N_d = \{1,2,\ldots,d\}$ denote the first d positive integers. For each subset $S \subset N_d$ with $|S| = s$, we let $\pi(X,S)$ denote the image of projecting X onto R^s by retaining the s specified coordinates of each point and discarding the other d-s coordinates. Using projections from R^3 to R^2 as our model, we call $\pi(X,S)$ a shadow of X. The number $p(X,S) = |\pi(X,S)|$ is called the size of this shadow. The original file management problem seeks to determine

$$p(n,d,s) = \min_{\substack{X \subset R^d \\ |X| = n}} \max_{\substack{S \subset N_d \\ |S| = s}} p(X,S),$$

that is, for each X we find a largest shadow size. We wish to determine the set X which minimizes this largest shadow size. Any set X achieving $p(n,d,s) = \max_{|S| = s} p(X,S)$ will be called an optimal set. Figure 1 provides a

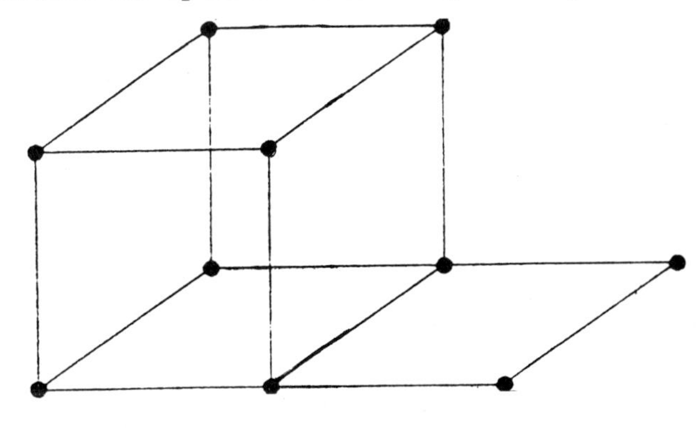

Figure 1. A set of ten points with largest shadow of size six.

depiction of 10 points in R^3 whose 3 projections to R^2 have sizes 4, 5, and 6. Their geometric means is $\sqrt{4\cdot 5\cdot 6} \cong 4.93$ while $10^{2/3} \cong 4.64$. This example shows that $p(10,3,2) \leq 6$ while the theorem below guarantees that $p(10,3,2) \geq 10^{2/3}$. Thus $p(10,3,2)$ is 5 or 6.

THEOREM 1. (Schwenk and Munro [3])
For any set X of n points in R^d, the projection size is bounded by

$$\text{Geom. Mean } \{p(X,S) : S \subset N_d \text{ and } |S| = s\} \geq n^{s/d}.$$

Furthermore, if n can be factored as $n = a_1 a_2 \ldots a_d$ with each a_i a natural number, then the bound is attained by choosing X to be the Cartesian product $\overset{d}{\underset{i=1}{\times}} N_{a_i}$.

Sketch of Proof. To prove this result in full requires multiple inductions and Hölder's Inequality for finite sets (Inequality (2.7.2) in Hardy, Littlewood, and Pólya [2, page 22]) and would duplicate the efforts in [3]. We shall present the two easiest cases to illustrate the flavor of the argument.

Case 1. For s = 1, and n and d arbitrary. This case is little more than the pigeonhole principle, for X is a subset of the Cartesian product $\overset{d}{\underset{i=1}{\times}} \pi(X,\{i\}) \supset X$.

Taking the order of both sides yields $\prod_{i=1}^{d} p(X,\{i\}) \geq n$.

Computing d th roots completes the proof.

Case 2. For d = 3, s = 2, and n arbitrary. Through each point of X we construct a horizontal plane and vertical line. Various points may generate the same plane or the same line. Figure 2 shows an example of 10 points

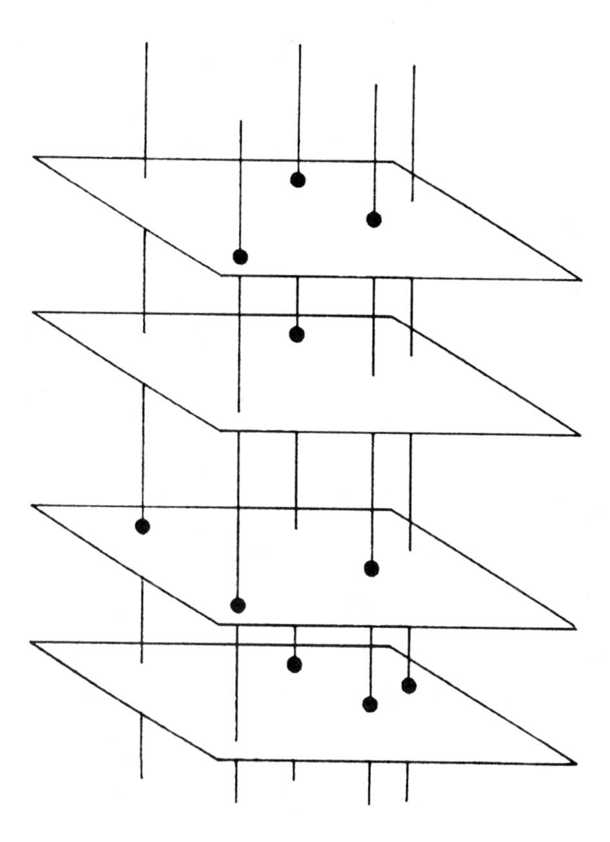

Figure 2. Ten points that generate four planes and five lines.

which generate 4 planes and 5 lines. Let us say there are $k \leq n$ planes and $r \leq n$ lines. Let X_i denote the subset of points of X in the i th plane and let $n_i = |X_i|$. Clearly $1 \leq n_i \leq r$ and $\sum_{i=1}^{k} n_i = n$. Let $x_i = p(X_i, \{1\})$ be the number of distinct x coordinates in plane i and let $y_i = p(X_i, \{2\})$ be the number of distinct y coordinates. Now $p(X, \{1,3\})$ is evidently the sum $\sum_{i=1}^{k} x_i$ and similarly $p(X, \{2,3\}) = \sum_{i=1}^{k} y_i$ while $p(X, \{1,2\}) = r$.

Thus $\prod_{|S|=2} p(X,S) = r(\sum_{i=1}^{k} x_i)(\sum_{i=1}^{k} y_i)$. The Cauchy-Schwarz inequality (which is an instance of Hölder's inequality with two factors) implies $\prod_{|S|=2} p(X,S) \geq r(\sum \sqrt{x_i y_i})^2$.
But the pigeonhole principle requires $x_i y_i \geq n_i$, so $\prod p(X,S) \geq r(\sum \sqrt{n_i})^2 = (\sum_{i=1}^{k} n_i \sqrt{r/n_i})^2$. But $r \geq n_i$ implies the radicals in the last expression are bounded by 1 so that $\prod_{|S|=2} p(X,S) \geq (\sum_{i=1}^{k} n_i)^2 = n^2$. Taking cube roots completes the proof.

The theorem provides that every set X satisfies $\max_{|S|=s} p(X,S) \geq \text{Geom Mean}\,\{p(X,S) : |S| = s\} \geq n^{s/d}$, so we know that $p(n,d,s) \geq \lceil n^{s/d} \rceil$. For $s = 1$ we set $k = \lceil n^{1/d} \rceil$. Let N_k^d denote the Cartesian product of d copies of N_k and let X be any subset of N_k^d with order n. Then $\max_{|S|=1} p(X,S) = k$ so $p(n,d,1) = \lceil n^{1/d} \rceil$ and X is optimal.
For larger values of s determining $p(n,d,s)$ becomes extremely difficult. We shall limit our attention to the first nontrivial case having $s = 2$ and $d = 3$.

2. COMPLETE SOLUTION FOR DIMENSION THREE

In this section we shall derive the precise value of $p(n,3,2)$ for each n and we shall present an optimal set attaining this minimum largest shadow size. The solution is expressed in terms of the parameters k and e where $k^3 \leq n = k^3 + e < (k+1)^3$. That is, k is the largest cube no exceeding n and e is the excess of n over k^3. Notice $0 \leq e \leq 3k^2 + 3k$.

Theorem 2. For $d = 3$, $s = 2$, $n = k^3 + e$, the smallest maximum projection size is given by

$$p(n,3,2) = \begin{cases} k^2 + \lceil\sqrt{e}\rceil & \text{for } e \le k^2 \\ k^2 + \lceil k/2 + \sqrt{e - 3k^2/4}\rceil & \text{for } e > k^2 \end{cases} .$$

We approach the theorem via two lemmas.

Lemma 3. There exists an optimal set X^* whose points have only natural numbers for coordinates with the additional property that $(a,b,c) \in X^*$ implies $N_a \times N_b \times N_c \subset X^*$.

Proof of Lemma. For each coordinate we may replace the distinct values that occur (say there are r of these) with the first r natural numbers N_r. This replacement does not alter any of the projection sizes. Now among all optimal $X \subset N^3$ for which $\max p(X,S) = p(n,3,2)$ select one X^* whose total of all coordinates for all n points is minimal. If $(a,b,c) \in X^*$, we claim $N_a \times N_b \times N_c \subset X^*$. If not, there exists $(x,y,z) \in N_a \times N_b \times N_c - X^*$. The four points (x,y,z), (x,y,c), (x,b,c), and (a,b,c) include a consecutive pair of points with the first not in X^*, the second in X^*, and two identical coordinates. For example, suppose $(x,y,c) \notin X^*$ and $(x,b,c) \in X^*$. Let gravity act in the direction of the y-coordinate causing points to slide down to fill in vacant lower integral positions producing a new set Y. Clearly this process does not alter the size of the x-z shadow while the x-y and y-z shadows can only decrease. That is, $p(n,3,2) \le \max p(Y,S) \le \max p(X^*,S) = p(n,3,2)$. But this means Y is another optimal subset of N^3 with total sum of coordinates smaller than X^*, we may conclude that $N_a \times N_b \times N_c \subset X^*$ whenever $(a,b,c) \in X^*$ as claimed.

Lemma 4. For $e \le k^2$ there exists an optimal X^* which is a subset of $N_k \times N_k \times N_{k+1}$.

Proof. First, there exists a set X with $n = k^3 + e$ points comprised of the cube N_k^3 union with e points chosen from $\{k+1\} \times N_m^2$ where $m = \lceil \sqrt{e} \rceil$. Evidently ma $p(X,S) = k^2 + \lceil \sqrt{e} \rceil \le k^2 + k$ so $p(n,3,2) \le k^2 + \lceil \sqrt{e} \rceil$ when $e \le k^2$. Select an optimal $X^* \subset N^3$ having minimum total sum of coordinates. Assume $X^* \subset N_a \times N_b \times N_c$ with $a + b + c$ minimal and $a \le b \le c$. If $a \le k - 1$, the n points of X^* lie in a planes so one of the projections has at least $n/a \ge k^3/(k-1) > k^2 + k + 1$ points. Since this contradicts $p(n,3,2) \le k^2 + \lceil \sqrt{e} \rceil \le k^2 + k$, we may assume $a \ge k$.

Next, if $k + 1 \le b \le c$, we examine the set X^* considering the $y - z$ plane as its base. The conditions of Lemma 3 assert that at each position (i,j) with $(i,j) \in N_b \times N_c$ there is a stack of integer lattice points $1,2,\ldots,h$ with height $h = h(i,j)$. Moreover, Lemma 3 implies that $h(1,1) \ge h(2,1) \ge \ldots \ge h(b,1)$ and similarly $h(1,1) \ge h(1,2) \ge \ldots \ge h(1,c)$.

We consider these positions as forming a "wall" of varying height comprising two sides of a rectangle behind which stacks of points may be hidden from the two sideways projections ($x - y$ and $x - z$) provided the stacks do not exceed the minimum wall height. This configuration is depicted in Figure 3. From the conditions given, this minimum height is either $h(b,1)$ or $h(1,c)$. For instance, if $h(b,1) \ge h(1,c)$ the minimum height is $h(1,c)$. Now $b(c-1) \ge (k+1)k \ge \max p(X^*,S)$ so the rectangle $N_b \times N_{c-1}$ must contain some empty positions lest $p(X^*,\{2,3\}) > p(n,3,2)$. In fact, there must be enough empty positions to accommodate the occupied

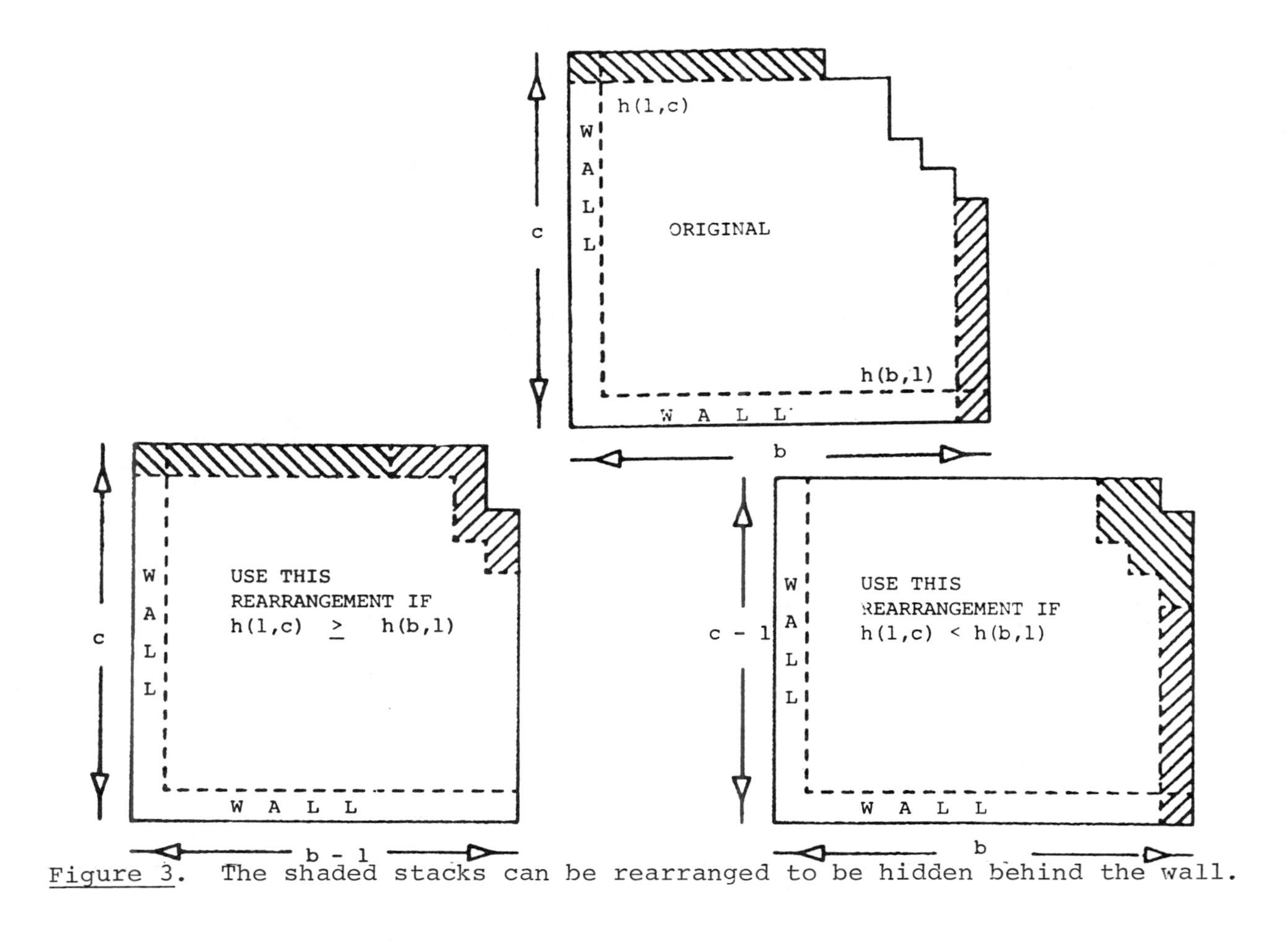

Figure 3. The shaded stacks can be rearranged to be hidden behind the wall.

columns from positions $(1,c), (2,c), \ldots, (b,c)$. These columns have heights $h(1,c) \geq h(2,c) \geq \ldots \geq h(b,c)$ so they all fit inside the rectangle $N_b \times N_{c-1}$ and are completely hidden from horizontal views by the wall. This rearrangement does not alter $p(X^*,\{2,3\})$ and $p(X^*,\{1,2\})$ while $p(X^*,\{1,3\})$ gets reduced by $h(1,c)$. Thus, $h(b,1) \geq h(1,c)$ implies an optimal set exists within $N_a \times N_b \times N_{c-1}$. Similarly, if $h(b,1) \leq h(1,c)$ we may produce an optimal set contained in $N_a \times N_{b-1} \times N_c$. Either case violates the minimality of $a + b + c$, so the supposition $b \geq k + 1$ leads to a contradiction. Consequently, $a = b = k$. Now the n points of X^* lie in c planes with at most $ab = k^2$ points in each plane. Let the ith plane have n_i points. Using the notation from the proof of Theorem 1 with $r = k^2$, Cauchy-Schwarz again implies $p(X^*,\{1,3\})\, p(X^*,\{2,3\}) = (\sum_{i=1}^{c} x_i)(\sum_{i=1}^{c} y_i)$ $\geq (\sum_{i=1}^{c} \sqrt{x_i y_i})^2$. The pigeonhole principle requires $x_i y_i \geq n_i$ so

$$p(X^*,\{1,3\})\, p(X^*,\{2,3\}) \geq \Big(\sum_{i=1}^{c} \sqrt{n_i}\Big)^2 = \Big(\frac{1}{k} \sum_{i=1}^{c} n_i \sqrt{k^2/n_i}\Big)^2.$$

For $n = k^3 + e$ the last sum is minimized by setting $n_i = k^2$ for $i \leq k$ and $n_c = n_{k+1} = e$ to obtain $p(X^*,\{1,3\})\, p(X^*,\{2,3\}) \geq (k^2 + \sqrt{e})^2$. Thus $p(n,3,2) = \max p(X^*,S) \geq k^2 + \lceil \sqrt{e} \rceil$, so we have proved $c = k + 1$ as claimed and moreover Theorem 2 has been verified for $e \leq k^2$.

Lemma 5. For $k^2 < e \leq 3k^2 + 3k$ there exists an optimal X^* which is a subset of N_{k+1}^3.

Proof. First we shall present a candidate for X^* and then, ultimately, show that it is optimal. Let t be the smallest integer such that $n \le k^2(k+1) + t(k+t)$. Our candidate for an optimal set X is comprised of any n point subset of $Y_t = N_k \times N_k \times N_{k+1} \cup N_k \times \{k+1\} \times N_t \cup \{k+1\} \times N_t \times N_t$. Each projection for Y_t gives $p(Y_t,S) = k^2 + k + t$, so $p(n,3,2) \le \max p(X,S) \le k^2 + k + t$. The choice of t requires $e \le k^2 + tk + t^2$ which can be solved to give $t \ge -k/2 + \sqrt{e - 3k^2/4}$. Since t is a positive integer, we get $p(n,3,2) \le k^2 + \lceil k/2 + \sqrt{e - 3k^2/4} \rceil$. Thus, any X^* we obtain must give a maximum shadow no larger than this.

Assume an optimal X^* is chosen to be a subset of $N_a \times N_b \times N_c$ with $a + b + c$ minimal and $a \le b \le c$. Could $a \le k$? If it were, the choice of t requires that $n > k^2(k+1) + (t-1)(k+t-1)$, and so

$$n/a > k^2 + k + t - 1 + (t-1)^2/k$$
$$\ge k^2 + k + t .$$

While X^* may fail to be a subset of N^3_{k+1}, if $a \le k$ we find that $\max p(X^*,S) = \max p(X,S)$ so $X \subset N^3_{k+1}$ satisfies all the required conditions.

We continue with the assumption $a \ge k + 1$. If $k + 2 \le b \le c$ we use the "wall" approach of Lemma 4 to reduce b or c. The rearrangement of columns of points must fit because both $b(c-1)$ and $c(b-1)$ are at least $(k+2)(k+1)$. If there were too many columns, the projection of X^* onto its base plane would exceed $(k+2)(k+1)$ whereas we already have a candidate $X \subset N^3_{k+1}$ for which $\max p(X,S) \le (k+1)^2$. Thus $b = a = k + 1$. Finally, if $c \ge k + 2$ the wall argument would allow a reduction unless the number of columns exceeds one of $b(c-1)$ or

$(b-1)c$. But $b(c-1) \geq (b-1)c \geq k(k+2) = k^2 + 2k$. This means the number of columns is at least $(k+1)^2$, so again, we may revert to our first set X as an optimal set. Thus $a = b = c = k + 1$ as required.

We have shown that $X^* \subset N^3_{k+1}$ and $\max p(X^*,S) \leq \max p(X,S) = k^2 + k + t$. We shall complete the proof of Theorem 2 by demonstrating that if $\max p(X^*,S) \leq k^2 + k + t - 1$, then n would be at most $k^3 + k^2 + (t-1)(k+t-1)$. But this contradicts the choice of t. To see this, note that $X^* \subset N^3_{k+1}$ and $\max p(X^*,S) \leq k^2 + k + t - 1$ implies each axial direction contains at least $(k+1)^2 - (k^2 + k + t - 1) = k + 2 - t$ vacant lines. Call this quantity $s = k + 2 - t$. Since $t \geq 1$, we have $s \leq k + 1$. Consider the $x - y$ plane as a base. The wall argument allows us to place all s vacant lines in one of two planes either $x = k + 1$ or $y = k + 1$. Without loss of generality suppose they lie in $x = k + 1$ This gives $s(k + 1)$ empty cells. Now let the $y - z$ plane be the base. The s vacant lines can be moved to either $y = k + 1$ or $z = k + 1$. In either case, each line overlaps with exactly one cell from the set in $x = k + 1$, so we have $s(k + 1 - 1)$ additional empty cells. Finally, the $x - z$ base provides empty lines parallel to the y axis. No matter where these lines are positioned, each has at most s cells in common with the vacant lines already counted. Thus we have a total of at least $s(k + 1) + sk + s(k + 1 - s)$ vacant cells in N^3_{k+1}. Substituting $s = k + 2 - t$ and carrying out routine algebraic manipulations yields $n \leq k^3 + k^2 + (t-1)(k+t-1)$. Since this violates the choice of t, we have shown that $p(n,3,2) = k^2 + k + t$. As noted above, the conditions on t yield $p(n,3,2) = k^2 + \lceil k/2 + \sqrt{e - 3k^2/4} \rceil$.

Theorem 2 assures us that $p(n,3,2) = n^{2/3}$ precisely when $n = k^3$ is a perfect cube. For values of n with nonzero excess e it is interesting to ask how much $p(n,3,2)$ might exceed $n^{2/3}$.

Theorem 6.

$$\limsup_{n \to \infty} \frac{p(n,3,2) - n^{2/3}}{n^{1/3}} = 3/8$$

and the maximum discrepancy of $3/8\ n^{1/3}$ is approached as $n \to \infty$ both for $e = 9k^2/16$ and for $e = 21\ k^2/16$.

Proof. For n (and hence k) sufficiently large, $n^{1/3} \cong k + e/3k^2$ and $n^{2/3} \cong k^2 + 2e/3k$. We consider the two ranges of excess separately. For $e \leq k^2$ we set $r = e/k^2 \leq 1$. Now

$$\frac{p(n,3,2) - n^{2/3}}{n^{1/3}} \cong \frac{k^2 + \sqrt{e} - (k^2 + 2e/3k)}{k + e/3k^2}$$

$$\cong \frac{\sqrt{r}k - 2rk/3}{k + r/3}$$

$$\cong \frac{3\sqrt{r} - 2r}{3 + r/k}$$

$$\cong \sqrt{r} - 2r/3 \; .$$

The last expression has its maximum value of 3/8 when $r = 9/16$ which means $e = 9k^2/16$.

On the other hand, for $e > k^2$ we set $r = e/k^2 - 3/4$. Thus $1/4 < r \leq 9/4$. In this range of excess, we find

$$\frac{p(n,3,2) - n^{2/3}}{n^{1/3}} \cong \frac{k^2 + k/2 + \sqrt{e - 3k^2/4} - (k^2+2e/3k)}{k + e/3k^2}$$

$$\cong \frac{k/2 + \sqrt{r}k - (2r/3 + 1/2)k}{k + r/3 + 1/4}$$

$$\cong \sqrt{r} - 2r/3 \; .$$

Since we have the same approximating expression as before, we approach the same maximum of 3/8 be selecting $r = 9/16$ which means $e = 21\,k^2/16$. This yields the rather surprising conclusion that for sufficiently large k the simple lower bound of $n^{2/3}$ is attained precisely when there is no excess while the more complicated expression $n^{2/3} + 3/8\,n^{1/3}$ is approached twice on every interval between consecutive perfect cubes, namely when $e = 9k^2/16$ and again when $e = 21\,k^2/16$.

The graph of $y = \dfrac{p(n,3,2) - n^{2/3}}{n^{1/3}}$ as a function of $x = e/k^2$ is rather interesting. For $0 \le x \le 1$ we have $y \cong \sqrt{x} - 2x/3$ while for $1 \le x \le 3$ we find that $y \cong \sqrt{x - 3/4} - 2(x - 3/4)/3$, the same curve translated 3/4 units to the right! Moreover, rotating the coordinate system by an angle $\theta = \arctan 1.5$ converts $y = \sqrt{x} - 2x/3$ to new variables yielding

$$\frac{-27}{13^{1.5}}\left(y' - \frac{3}{13^{1.5}}\right) = \left(x' - \frac{9}{13^{1.5}}\right)^2 .$$

Thus, we recognize the curve as a parabola with vertex (in the original coordinates at (9/169, 33/169) and axis of symmetry given by $y = 3/13 - 2x/3$. A sketch of this parabola and its translate is shown in Figure 4. Note the double attainment of the maximum for y at $x = 9/16$ and 21/16.

3. THE HIGHER DIMENSIONAL PROBLEM

The evaluation of $p(n,d,s)$ for $d \ge 4$ appears to be very difficult. The lower bound of $\lceil n^{s/d} \rceil$ from Theorem 1 still applies and suggests the general size of the answer. Lemma 3 generalizes to prove an integral optimal

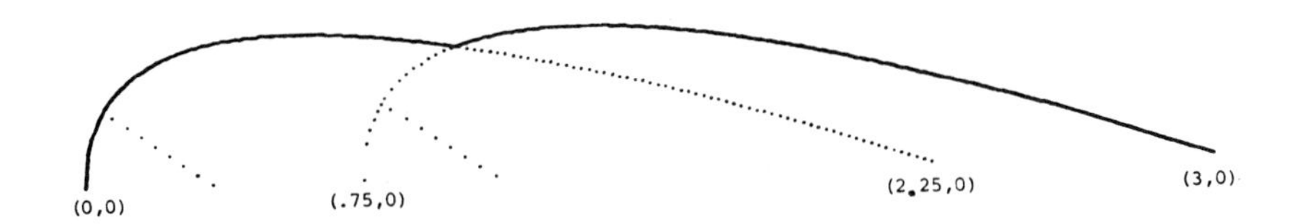

Figure 4. A parabola and its translate.

set $X \subset N^d$ exists and $(a_1, a_2, \ldots, a_d) \in X$ implies $\mathop{X}_{i=1}^{d} N_{a_i} \subset X$. However, the wall argument used to prove $X \subset N^3_{k+1}$ in Lemmas 4 and 5 breaks down for $d \geq 4$. If n happens to have a very special form, we are able to provide an upper bound for p(n,d,d-1).

Lemma 7. If $n = \sum_{i=1}^{d} a_i^i$ with $0 \leq a_1 \leq a_2 \leq \cdots \leq a_d$, then $p(n,d,d-1) \leq \sum_{i=1}^{d} a_i^{i-1}$.

Proof. The upper bound is attained by the set $X = \bigcup_{i=1}^{d} N^i_{a_i} \times \{1 + a_d\}^{d-i}$. The projection that suppresses the first coordinate produces $\sum_{i=1}^{d} a_i^{i-1}$ while every other projection is smaller.

We conjecture that this upper bound is the correct value for p(n,d,d-1) for those values of n that can be represented by $\sum_{i=1}^{d} a_i^i$ with $0 \leq a_1 \leq a_2 \leq \cdots \leq a_d$.

REFERENCES

1. H. Alt, K. Mehlhorn, and J. I. Munro, Partial match retrieval in implicit data structures, to appear.
2. G. H. Hardy, J. E. Littlewood, and G. Pólya, Inequalities, Cambridge University Press, London, 1967.
3. A. J. Schwenk and J. I. Munro, How small can the mean shadow of a set be?, _Amer. Math. Monthly_ 253 (1983), 325-329.

THE OCHROMATIC NUMBER OF PLANAR GRAPHS[†]

Gustavus J. Simmons

Sandia National Laboratories
Albuquerque, New Mexico 87185

ABSTRACT. Given an ordering ϕ of the vertices of a graph G, let $\chi_\phi(G)$ denote the least number of colors that suffice to color the vertices in the order ϕ, subject to the rule that a new color is introduced only when a vertex of the graph cannot be properly colored in its order with any of the colors already used. If the coloring is not unique, i.e., if for some vertex a choice exists among a subset of the colors already used, then $\chi_\phi(G)$ is taken to be the minimum number of colors that will suffice for at least one sequence of choices. The maximum of $\chi_\phi(G)$ over all possible orderings of the vertices of G is defined to be the ochromatic number of G denoted by $\chi^\circ(G)$. Surprisingly even for trees where $\chi(T) = 2$, $\chi^\circ(T)$ is unbounded as a function of the number of vertices. A natural question is to determine the minimum number N_k of vertices, a k-ochromatic planar graph can have. Upper and lower bounds on N_k are given.

[†]This work performed at Sandia National Laboratories supported by the U.S. Department of Energy under contract number DE-AC04-76DP00789.

1. INTRODUCTION

The ochromatic number of a graph, $\chi^{\circ}(G)$, was first introduced by the author in [4] where it was called the ordered chromatic number--with the same notation. In view of the novelty of this graph parameter, a few remarks about the motivation for its introduction seem appropriate.

In the usual definition of the chromatic number $\chi(G)$ as the minimum number of colors with which the vertices of the graph G can be properly colored, there is no explicit indication of any dependence on an ordering of the vertices. However, if one wishes to either approximate, or determine exactly, the chromatic number of a nontrivial graph, almost all algorithms proceed by sequentially assigning colors to the vertices--generally in accordance with a local decision rule, such as the greedy algorithm, etc. Often back tracking is done after each coloring is completed to alter the order in which the vertices were colored or the choice of permissible proper colors, where choices were made, so as to decrease the total number of colors required. There is a large literature, consisting on the one hand of various local decision rules to "efficiently" properly color graphs and on the other of constructions of graphs on which these rules fail miserably. It is inappropriate to this paper to do more than mention this obliquely related literature which was surveyed by Matula, Marble and Isaacson [3], however, the initial motivation for the ochromatic number came from the study of ordered vertex colorings of graphs. They note that for every graph G there is at least one ordering of the vertices for which a local decision rule that they call a <u>sequential coloring</u> uses only $\chi(G)$ colors. They use the graph in Figure 1 to demonstrate the dependence of the number of colors required on the ordering of the vertices.

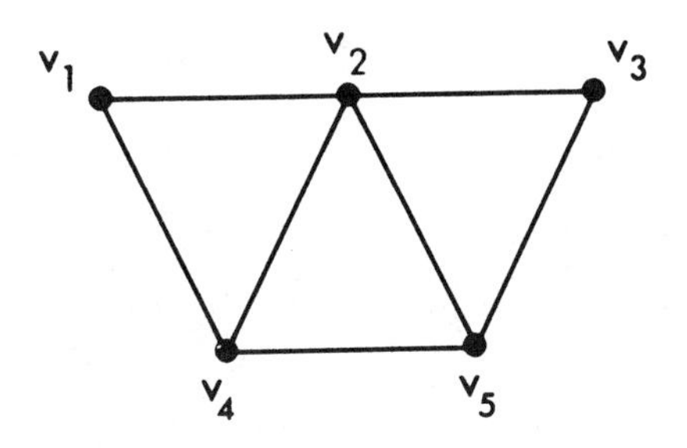

Figure 1.

For the order $\phi_1 = (v_1,v_2,v_3,v_4,v_5)$ four colors are needed, while for $\phi_2 = (v_5,v_4,v_3,v_2,v_1)$ three suffice. As it happens the graph of Figure 1 can also be used to illustrate all of the points in the definition of the ochromatic number.

Given an ordering ϕ of the vertices of a graph G, let $\chi_\phi(G)$ denote the least number of colors that a parsimonious painter, i.e., one who only opens a new tube of paint when a vertex of the graph cannot be properly colored in its order with any of the tubes already opened--would need in his palette to properly color the vertices of G in order ϕ. Such a coloring we will call a <u>parsimonious proper coloring</u> (PPC). If the coloring is not unique, i.e., if for some vertex or vertices a choice exists among a subset of the colors already used, then $\chi_\phi(G)$ is taken to be the minimum number of colors that will suffice for at least one sequence of choices.

The minimum of $\chi_\phi(G)$ over all possible orderings, ϕ, of the vertices is simply the <u>chromatic number</u> of G,

2. MINIMAL k-OCHROMATIC PLANAR MAPS

We first recursively construct a family of maps m_p, whose dual graphs are trees, and define orderings of the regions for which the ochromatic number of the k'th member is k. The first four maps are:

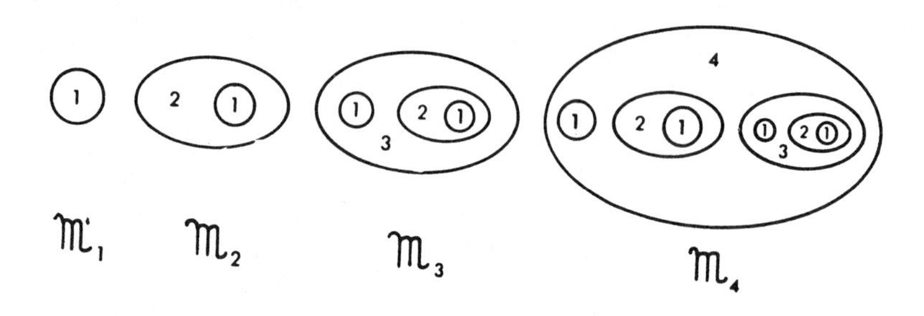

Figure 2

The extension to other maps is obvious: to construct m_k replicate each of the lower order maps inside a new containing region preserving the inherited color classes. Clearly, any ordering that first exhausts all of the regions in class 1 and then in class 2, etc., will result in $\chi^\circ(m_k) = k$ as claimed. One of the reasons for exhibiting this example is to illustrate this device for defining vertex orderings in recursive constructions which we shall use throughout the rest of this paper--the other is to make the transition from planar graphs to their dual maps which will be much more convenient for our discussion.

Since trees are both bipartite and planar this construction shows that no finite number of colors can suffice for the ordered coloring of all members of

written $\chi(G)$ as usual. The maximum of $\chi_\phi(G)$ over all possible orderings, ϕ of the vertices we define to be the <u>ochromatic number</u> of G, denoted by $\chi^\circ(G)$.

Referring to the graph of Figure 1 and the vertex orderings ϕ_1 and ϕ_2:

$$\chi_{\phi_1}(G) = \chi(G) = 3$$

$$\chi_{\phi_2}(G) = 4$$

with no choice of color assignment at any vertex in either ordering for a PPC. Consider now the ordering $\phi_3 = (v_1, v_4, v_3, v_2, v_5)$. Here v_1 is assigned color 1 and v_4 must be assigned color 2. Since v_3 is adjacent to neither v_1 nor v_4, it could be colored with either color 1 or color 2. If color 1 is selected, four colors are required, while if color 2 is selected only three are needed. Since $\chi(G) = 3$ at least three are needed, therefore by the definition of $\chi_\phi(G)$;

$$\chi_{\phi_3}(G) = 3 .$$

In fact $\chi^\circ(G) = 4$, so that the ordering ϕ_2 of [3] to illustrate a point about sequential colorings is also an illustration of the 4-ochromatic coloring of their graph.

In [4] it was shown that while the chromatic number of bipartite graphs is only two and of planar graphs is at most four, no finite number of colors suffice for a PPC of all trees--nor hence of bipartite or planar graphs either. The question investigated there and pursued at length here is to bound the minimum number of regions a planar map can have for which $\chi^\circ(G) = k$ where G is the dual graph to the map.

either of these classes of graphs. In view of this result, the obvious question is:

<u>Question:</u> What is the least number of regions, N_k, in a planar map with ochromatic number k?

The balance of this paper is devoted to determining bounds on these minimal k-ochromatic planar maps.

Since the number of regions in m_k is 2^{k-1}, the construction just given shows that

$$N_k \leq 2^{k-1} . \qquad (1)$$

We next construct two families of planar maps in much the same way that the maps m_i were constructed--only now the outer boundary of each k-ochromatic map will be divided between color classes k and k-1. We distinguish between the cases; k odd and k even. Figure 3 shows the first three members of the construction for k even. The continuation is evident and results in an m_k map having $(\sqrt{2})^k$ regions.

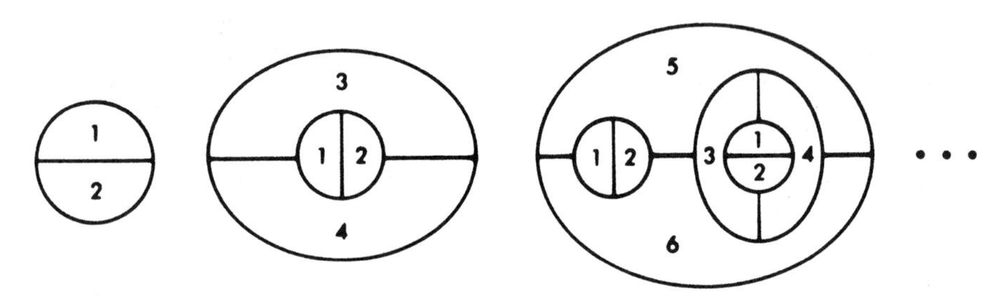

Figure 3

Figure 4 shows the first four maps of the recursive construction for k odd. For k = 2m-1 > 7 one replicates the minimal maps for k = 3 and 5 and the recursively generated maps for k = 7,9,..., 2m-3. The number of regions in the resulting k-ochromatic maps is $11(\sqrt{2})^{k-7}$.

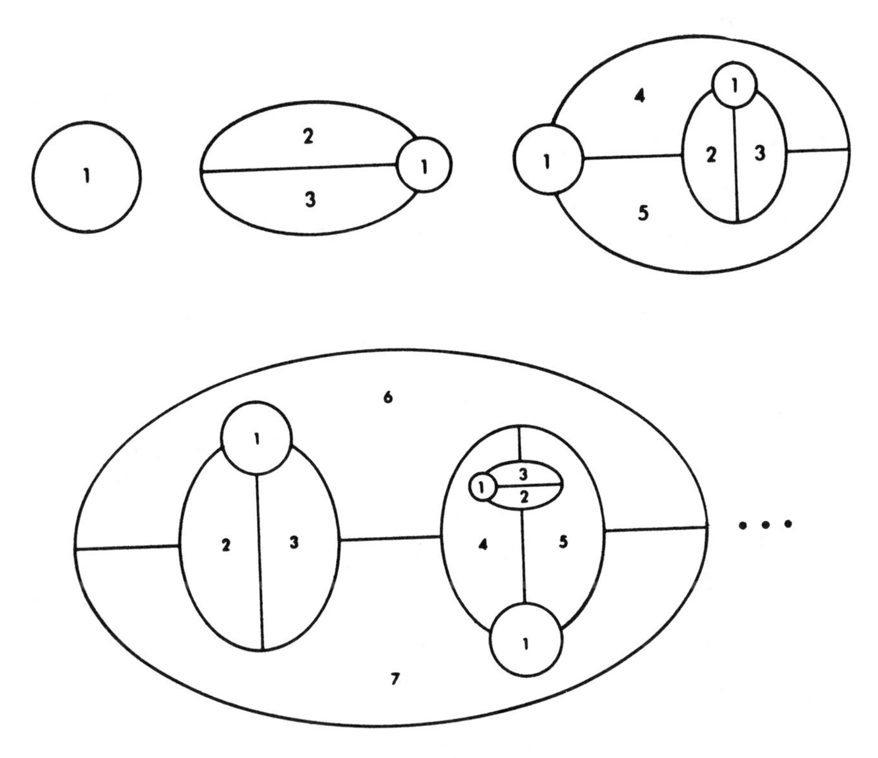

Figure 4

Since

$$11(\sqrt{2})^{k-7} < (\sqrt{2})^k$$

for $k \geq 7$, this construction and the one in Figure 4 shows that

$$N_k \leq (\sqrt{2})^k \qquad (2)$$

for both odd and even $k \geq 7$.

It is easy to show that N_k is at least quadratic in k. We derive the result for planar graphs with obvious duality for planar maps. As in [2], let $p = |V(G)|$ and $q = |E(G)|$.

Theorem 1.

$$N_k \geq \left\lceil \frac{k^2-k+12}{6} \right\rceil \tag{3}$$

Proof. Let G be a graph for which $\chi^o(G) = k$ and $p = N_k$. Consider the partition of p induced by any ordering for which $\chi_\phi(G) = k$. Since the coloring is a PPC, the first vertex assigned color i, for each i, must by definition be adjacent to at least one vertex in every lower order color class. The number q of edges in G is therefore at least $k(k-1)/2$. On the other hand, it is well known [2, p. 104] that the maximum number of edges a planar graph on p vertices can have is 3p-6. Therefore

$$\frac{k(k-1)}{2} \leq q \leq 3p-6 = 3N_k - 6 \tag{4}$$

and upon solving for N_k, the inequality follows.

We now use this lower bound, and simple arguments to obtain examples of minimal k-ochromatic graphs for $k \leq 8$. For $k \leq 4$, K_k is planar, Figure 5, and is the unique minimal k-ochromatic graph.

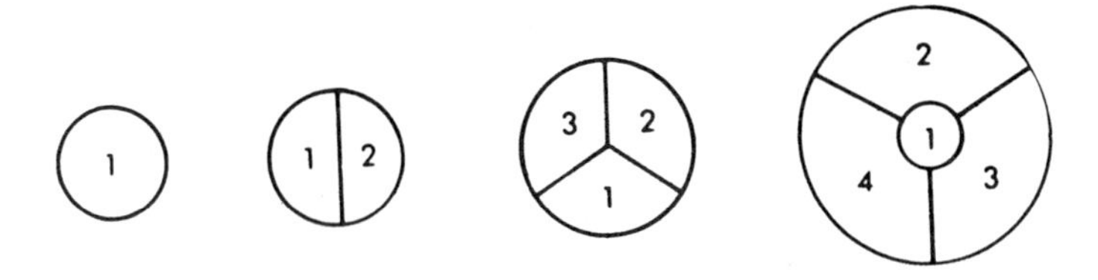

Figure 5

By Theorem 1, $N_5 \geq 6$. The map and the ordering of the regions shown in Figure 6 show that $N_5 = 6$. A thorough examination of the case k = 5 suggests techniques which can be used for the next few values of k. Since

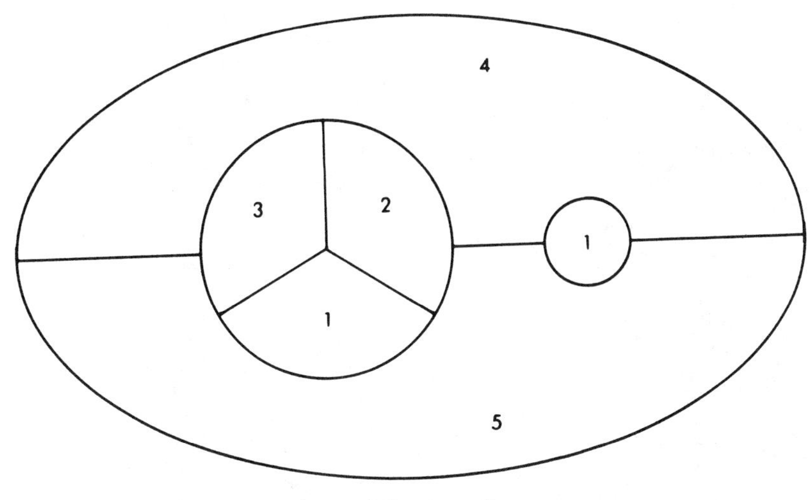

Figure 6

there are only six vertices in the minimal dual graph for k = 6 which are to be partitioned into five non-empty classes, one class contains precisely two vertices. These two vertices are nonadjacent since they are in the same color class under some ordering ϕ , therefore at most one color would be needed to color them if the other vertices were colored first--although they may not be colored the same as under ϕ. But when the other four vertices must define a graph G; for which $\chi°(G') = 4$. This is uniquely K_4, and hence the two vertices are forced to be in a single color class--with every vertex of K_4 adjacent to at least one of the pair. This can happen in exactly four nonisomorphic ways as shown in Figure 7. From (4), $q \geq 10$ so that the first two constructions not only minimize p, but also q for a 5-ochromatic graph. Figure 8 shows the corresponding planar maps--where the unlabeled regions (vertices respectively) can be colored in arbitrary order after color class 1.

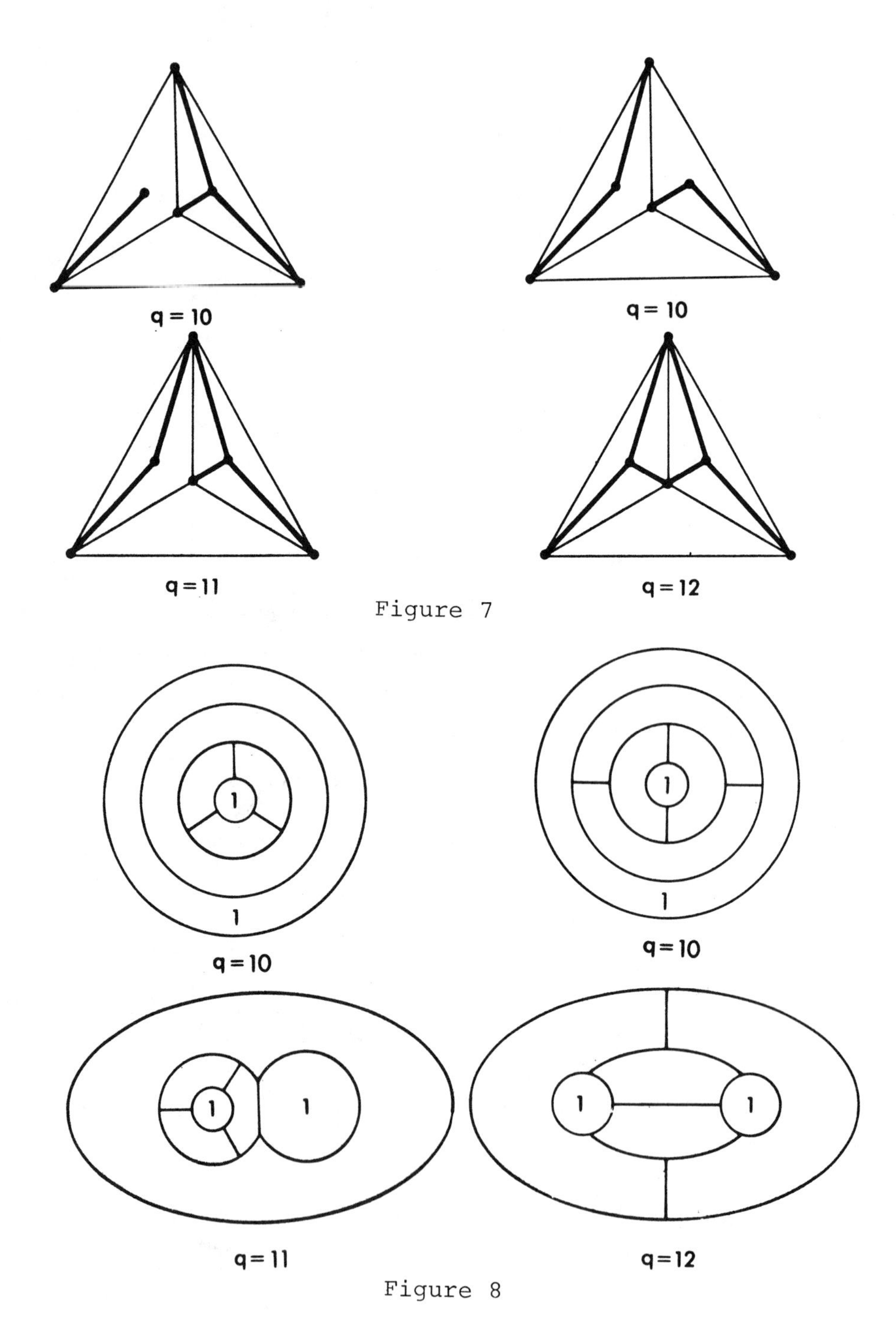

Figure 7

Figure 8

Theorem 1 states that $N_6 \geq 7$, $N_7 \geq 9$ and $N_8 \geq 12$. Using arguments similar to those just used for k = 5, it is easy to show by removing two vertices that are in the same color class, properly coloring the resulting subgraph, and then returning the missing vertices, that strict inequality holds for k = 6 and 7. Figures 9 and 10 show by construction that $N_6 = 8$ and $N_7 = 10$. The construction of Figure 11 demonstrates that $N_8 \leq 14$ and by easy arguments of the sort used above, $13 \leq N_8$. Unfortunately the argument only proves $13 \leq N_8 \leq 14$. We conjecture $N_8 = 14$.

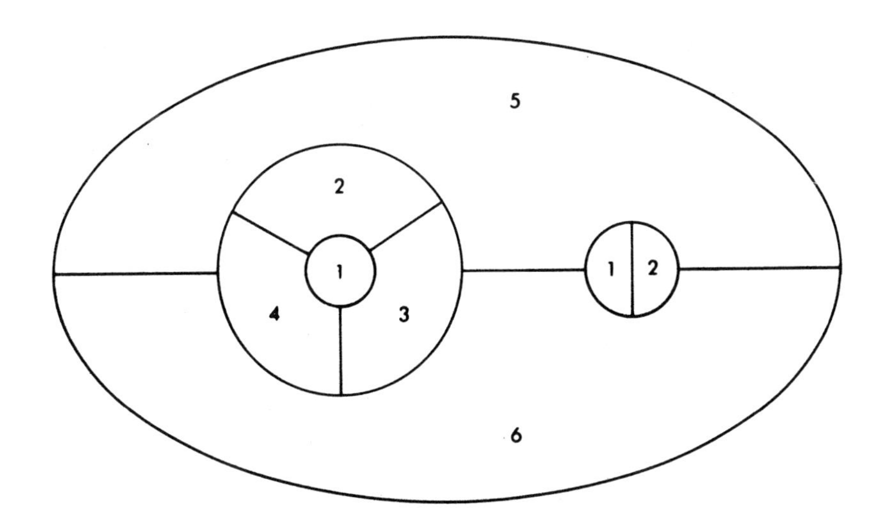

Figure 9

The trick used earlier to reduce the number of regions in a map requiring $k = 2m - 1$ colors was to "bud" an interior region of the map through the boundary to create a new adjacency and hence reduce the total number of regions needed in a color class. We now formalize this <u>ad hoc</u> procedure and systematically exploit it to get the best known upper bounds for N_k.

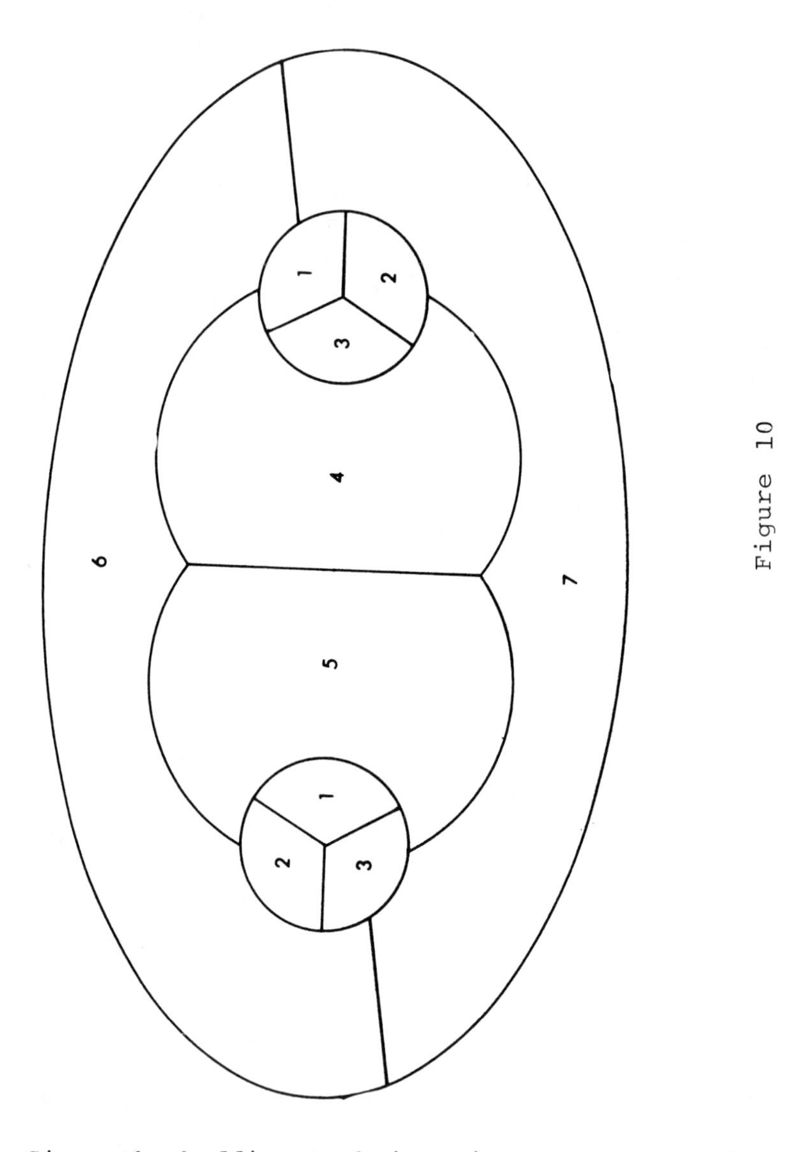

Figure 10

Since the budding technique is so easy to understand from an illustration and so cumbersome to describe, the reader is referred to Figure 12 showing two representative cases. There are two points to be made. First, when an interior j-ochromatic map is budded, either the

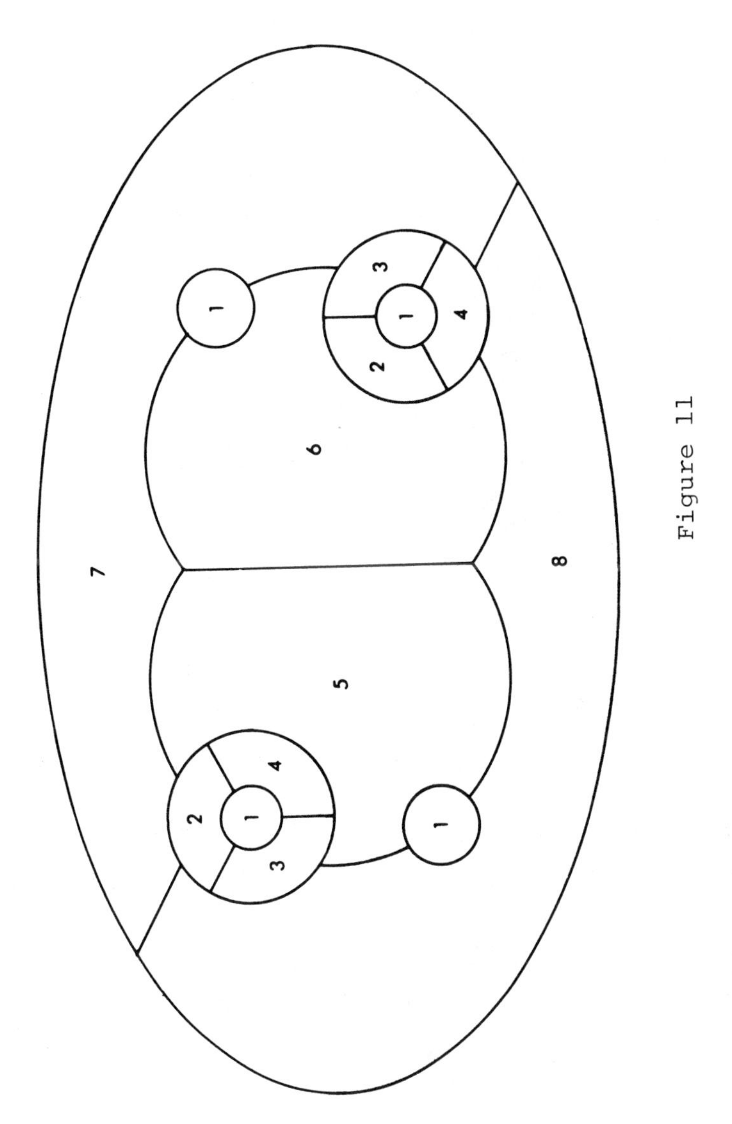

Figure 11

j or j-1 region can be caused to protrude through the boundary. Second, no matter how many maps have been replicated on the equator of the recursively defined k-ochromatic graph, only two can be budded through the boundary. It is true, and nearly obvious, that the most

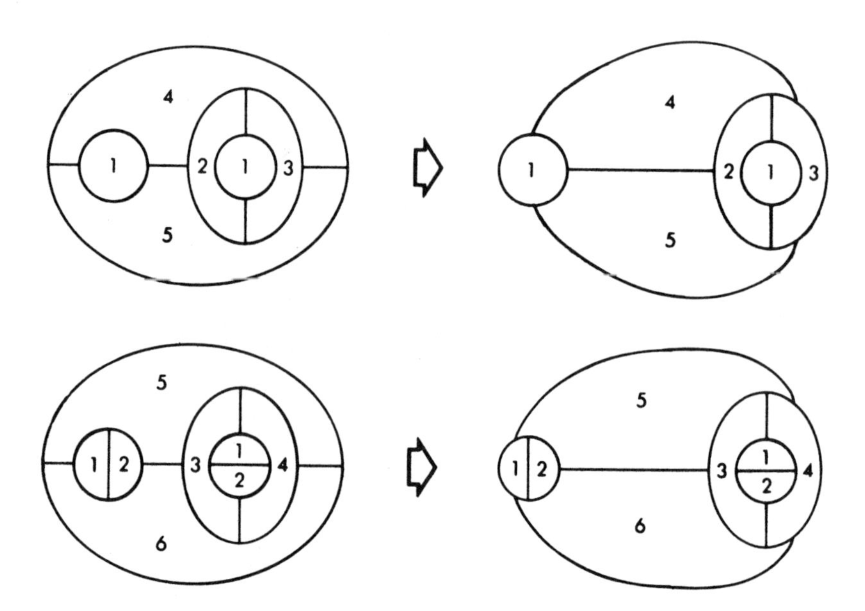

Figure 12

efficient use of the two buds is to bring the highest order colors possible to the outside of the map boundary.

Since even for small values of k the full maps are too complex to easily see what is happening, we introduce the suggestive notation for budded maps shown in Figure 13. A bud is said to be <u>free</u> if the construction does not require either of the color classes that it can provide to be budded in order to force the overall map to be k-ochromatic. The notation $\frac{i}{i-1}$ means that either color class i or color class i-1 or neither can protrude through the boundary of the bead without affecting the ochromatic number of the containing graph. Figure 14 shows the constructions corresponding to those shown in Figure 12 using this notation.

Notation

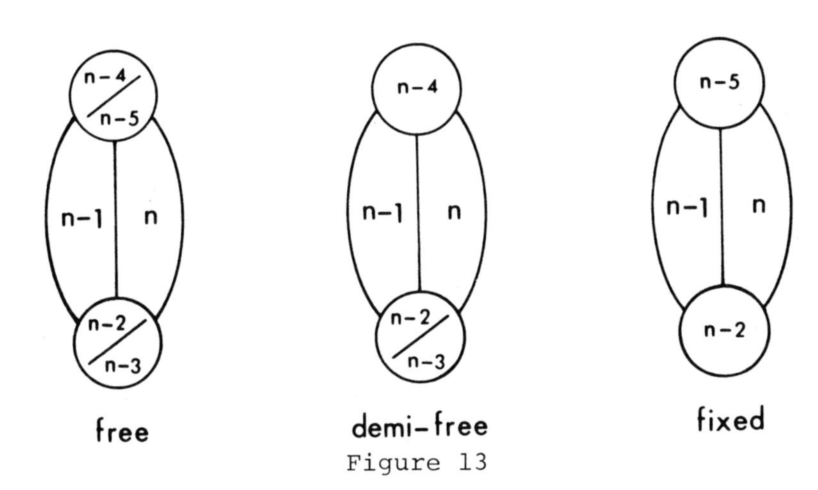

Figure 13

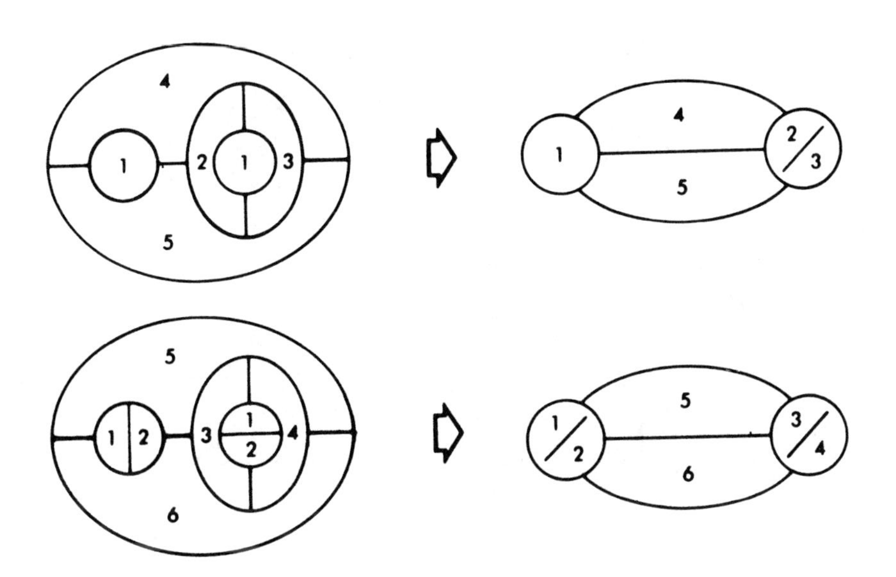

Figure 14

Using this notation for beads, Figure 15 shows the general recursive constructions for these cases.

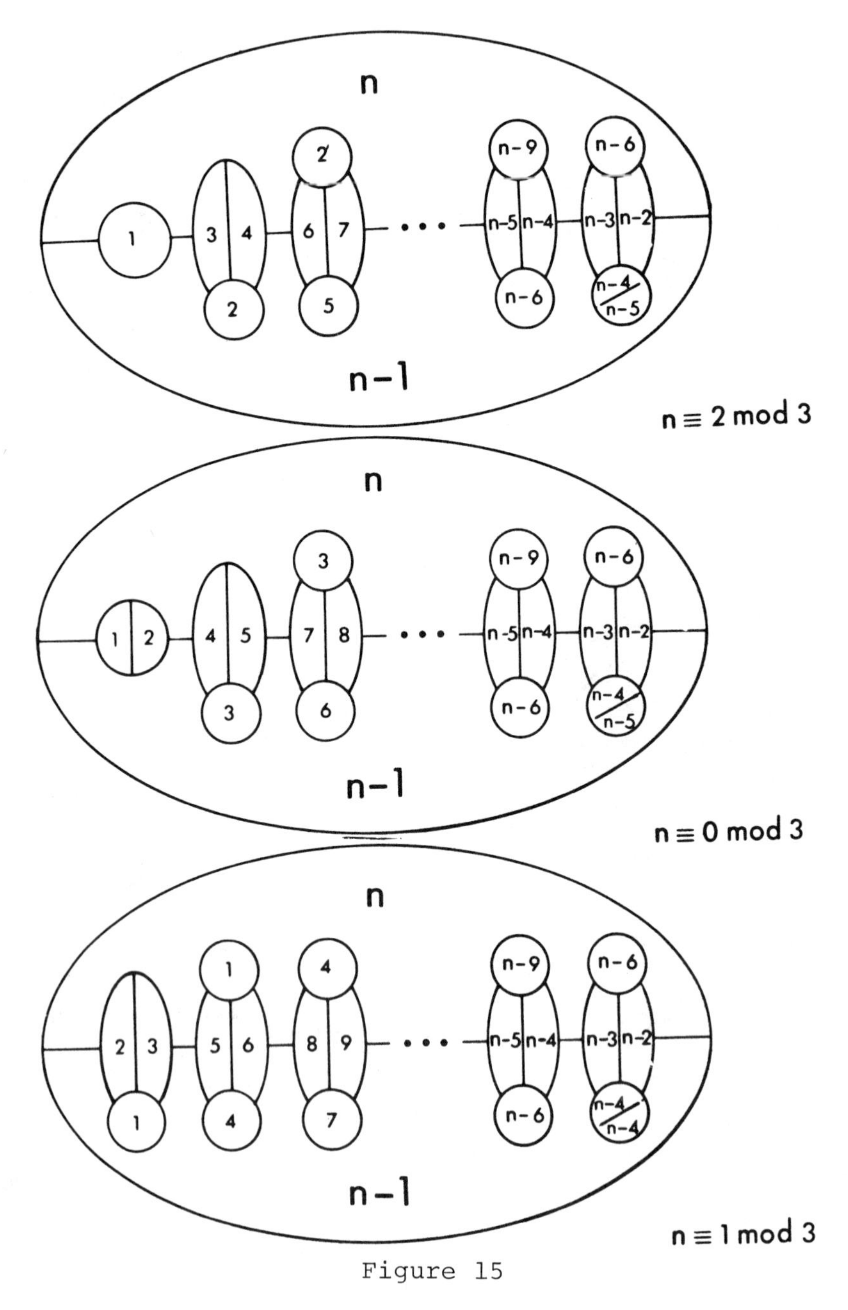

Figure 15

It is now a simple matter to tally the number of regions in these maps. Let F_i denote the number of regions in the i-ochromatic map. Then

$$F_k = 2 + F_{k-2} + \sum_{j=1}^{\left\lceil \frac{k-4}{3} \right\rceil} F_{k-4-3j} \tag{5}$$

where the constant 2 tallies the new outer regions in the map and the F_i are the region counts for the beads as shown in Figure 15. Subtracting F_{k-3} from F_k, we get the recursion

$$F_k - F_{k-2} - F_{k-3} - F_{k-4} + F_{k-5} = 0 \tag{6}$$

with the boundary conditions $F_i = N_i$; $i \leq 5$, i.e., precisely the anamalous cases mentioned earlier. Table 1 summarizes the results of this construction for $k \leq 16$.

Table 1

k	1	2	3	4	5	6	7	8	9	10	11	12	13	14	15	16
F_k	1	2	3	4	6	8	11	15	21	28	39	53	73	99	137	186

The limiting behavior of F_k is exponential of the form r^k where r is the root of

$$x^5 - x^3 - x^2 - x+1 = 0 \tag{7}$$

greater than 1:

$$r \approx 1.3690369436\ldots . \tag{8}$$

Unfortunately r^k approaches F_k from below so although we have dramatically improved the earlier bound of $N_k < (\sqrt{2})^k$ for $k \geq 7$, we only have an asymptotic (with k) limit for N_k.

There is still a little more to be gained from the budding technique. In Figure 15 we began the recursive construction with an 8-ochromatic map on fifteen regions so that the three families of maps could be illustrated in complete generality. The map for k = 7 is shown in Figure 16 in which we note that if the labels on the

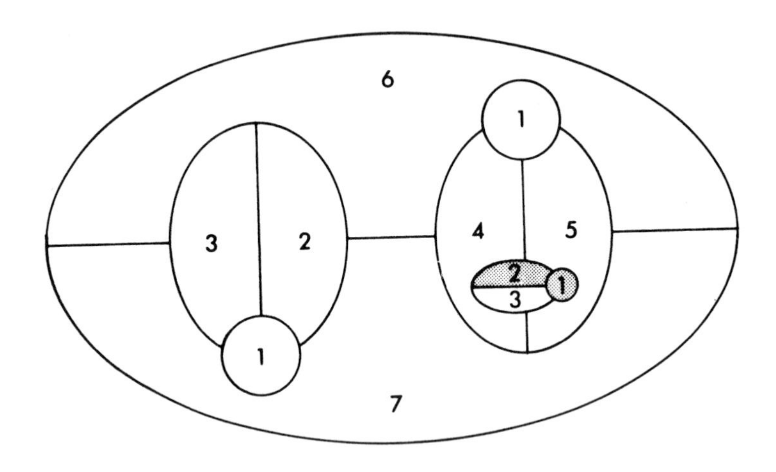

Figure 16

two shaded regions assigned to color classes 1 and 2 are interchanged, that the only essential adjacency destroyed is the one between the region assigned to color class 4 to a region in color class 2. However, the left and right beads can be pushed together along the equator to create a new class 4 to class 2 adjacency. The reason for doing this is that there are now two regions assigned to color class 1 in the right hand bead which can be joined by an isthmus to form a single region, Figure 17, thereby eliminating one region from the map. Incidentally, since this results in a 7-ochromatic map on ten regions, we know from the earlier argument that the resulting map is also minimal. There is a very

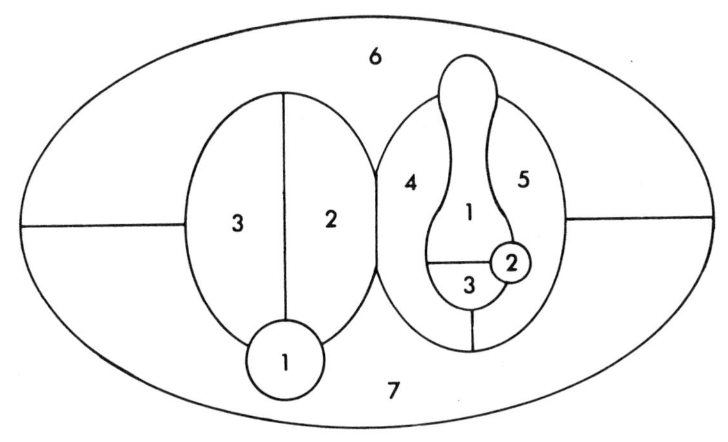

Figure 17

important point about such a reduction--since the two beads are now a Siamese pair, the resulting map can only be used as a fixed or demi-free bead in other constructions since the required adjacency between the regions in classes 6 and 7 prevents two buds from being formed.

As we have already remarked the bead construction gave minimal graphs for $k \leq 6$, hence the contraction just described is first possible for $k = 7$--with a resulting minimal construction. The same sort of contraction is also possible on the bead construction for $k = 8$ which reduces the region count from 15 to 14--reinforcing our earlier conjecture that $13 \leq N_8 \leq 14$ is actually $N_8 = 14$.

The number of regions that can be eliminated by contractions of this sort depends on whether a bead is fixed or else free or demi-free. The first fixed bead in which a reduction (of one region) is possible is for $k = 9$ when the demi-free interior bead of a 7-ochromatic map contributes one region to the contraction.

Represent an i-ochromatic demi-free bead by B_i and

an i-ochromatic fixed bead by b_i, then for all k, both B_k and b_k contain the demi-free bead B_{k-2} and the fixed beads

$$b_{k-4-3i} \quad \text{for} \quad 0 \le i \le \left\lceil \frac{k-4}{3} \right\rceil.$$

If we represent by R_i and r_i the number of regions that can be eliminated by the contractions in B_i and b_i, respectively, then

$$r_k = R_{k-2} + \sum_{i=0}^{\left\lceil \frac{k-4}{3} \right\rceil} r_{k-4-3i} \,. \tag{9}$$

In B_k it is possible to eliminate a single additional region by moving the demi-free (in bead B_k) bead B_{k-2} and the fixed bead b_{k-4} inside the boundary of B_k if and only if

$$k \equiv 1 \text{ or } 2 \pmod 6.$$

Therefore;

$$R_k = r_k + \left\lceil \frac{k-1}{6} \right\rceil - \left\lceil \frac{k-3}{6} \right\rceil. \tag{10}$$

Table 2 shows the number of regions that can be eliminated by these contractions for $k \le 16$.

Table 2

k	1	2	3	4	5	6	7	8	9	10	11	12	13	14	15	16
R_k	0	0	0	0	0	0	1	1	1	1	1	1	3	3	4	5
r_k	0	0	0	0	0	0	0	0	1	1	1	1	2	2	4	5

Combining (9) and (10) it is possible to express R_k by the recursion:

$$\begin{aligned} R_k - R_{k-2} - R_{k-3} - R_{k-4} + R_{k-5} \\ = 2\left\lceil \frac{k-1}{6} \right\rceil - \left\lceil \frac{k-3}{6} \right\rceil - \left\lceil \frac{k-4}{3} \right\rceil - \left\lceil \frac{k-5}{6} \right\rceil + \left\lceil \frac{k-6}{6} \right\rceil + 1 \end{aligned} \tag{11}$$

or more simply

$$R_k - R_{k-2} - R_{k-3} - R_{k-4} + R_{k-5} = f_k \qquad (12a)$$

where the forcing function f_k is given by

$$f_k = \begin{cases} 1 \\ -1 \\ -1 \\ 0 \\ 1 \\ 2 \end{cases} \quad \text{if} \quad k \equiv \begin{cases} 0 \\ 1 \\ 2 \\ 3 \\ 4 \\ 5 \end{cases} \pmod 6. \qquad (12b)$$

The most important thing to observe is that the homogeneous part of (12a) is the already familiar equation (7) so that the asymptotic behavior of the number of regions in the planar k-ochromatic maps recursively constructed by this procedure is still r^n as found before.

Table 3 summarizes the state of our knowledge at the moment about N_k. Unfortunately, although this improves the region count it does not get it below r^k, so the best we have shown is an asymptotic limit on N_k. N_k is almost certainly an exponential function s^k but where s lies in the interval 1 to 1.369...is an open and apparently hard question.

Table 3

k	1	2	3	4	5	6	7	8	9	10	11	12	13	14	15	16
N_k	1	2	3	4	6	8	10	14 (?)	≤18	≤24	≤32	≤52	≤70	≤96	≤133	≤181

REFERENCES

1. R. L. Brooks, On colouring the nodes of a network. Proc. Cambridge Philos. Soc. 37 (1941) 194-197.

2. F. Harary, Graph Theory. Addison-Wesley, Reading, MA (1969) 104.

3. E. W. Matula, G. Marble and J. D. Isaacson, Graph coloring algorithms. Graph Theory and Computing (R. C. Read, ed.) Academic, New York (1972) 109-122.

4. G. J. Simmons, The ordered chromatic number of planar maps. Proc. Thirteenth S.E. Conf. on Combinatorics, Graph Theory and Computing (1983), to appear.

A STUDY OF SNARK EMBEDDINGS

Frederick C. Tinsley
John J. Watkins

The Colorado College
Colorado Springs, CO 80903

ABSTRACT. A snark is a cubic graph with chromatic index 4. We investigate the genus of snarks and determine it for Isaacs' infinite family of flower snarks. Upper bounds are obtained for several other snarks. We also discuss the genus of a dot product of snarks.

1. INTRODUCTION

Interest in 3-regular graphs whose edges cannot be colored with three colors can be traced back to Tait's attempt [6] to prove the Four Color Theorem. Because of this theorem, we know that all such graphs are non-planar. Since they are so difficult to find, Martin Gardner [3] has named them 'snarks' after Lewis Carroll's "The Hunting of the Snark". The present state of snark hunting is well documented in [2]. In order to avoid trivial cases we make the following definition.

A snark is a cubic graph with chromatic index 4 which is cyclically 4-edge-connected and has girth at least 5.

It can be shown that there are arbitrarily large cubic planar graphs. In particular, there are cubic planar graphs with p vertices whenever p is a multiple of 4, for example, the graph in Figure 1.

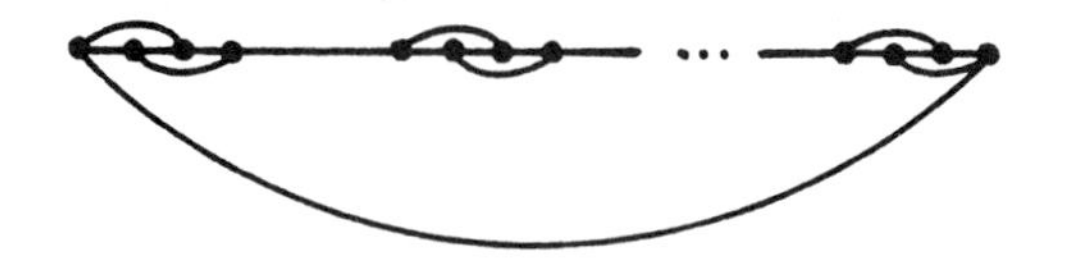

Figure 1.

This seems not to be the case with snarks. For the snarks we have investigated, the genus increases rather quickly with the order p of the graph. In fact, for the flower snarks the genus is given by $\gamma = 1/8\, p - 1/2$.

We can use a standard technique to find the genus of snarks of small order. If a cubic graph has p vertices and q edges, then $3p/2 = q$. Since snarks have girth at least 5, if we embed a snark of genus γ on a surface of genus γ with r regions, then $5r/2 \leqq q$. By applying Euler's formula, $2 - 2\gamma = p - q + r$, we easily get $\gamma \geqq 1 - p/20$. We conclude that for snarks with fewer than 20 vertices, the genus is greater than 0. Thus, the bound given by Euler's formula is very weak.

The smallest snark is the Petersen graph which has 10 vertices. Sinces it can be embedded on a torus, its genus is 1. We show it in Figure 2 in the form illustrated by Kempe [4].

The next smallest snark has 18 vertices and was discovered by Blanuša [1]. We draw it in the form shown in Figure 3. We can embed it on the torus to show that its genus is also 1, as in Figure 4.

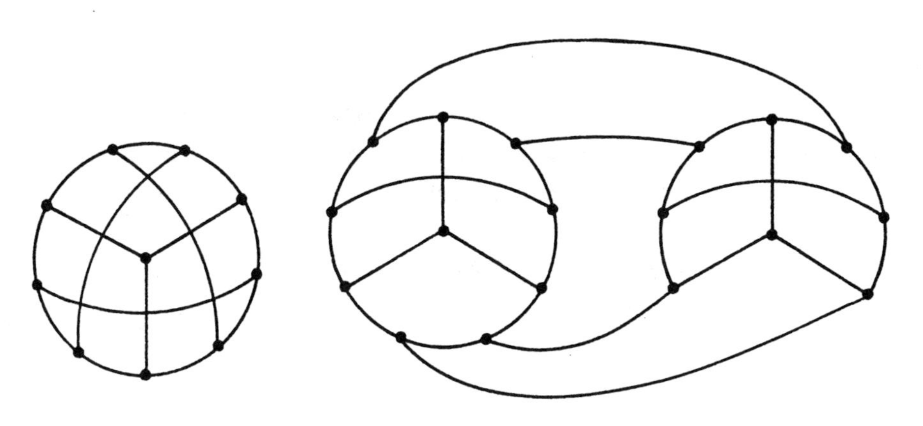

Figure 2.

Figure 3.

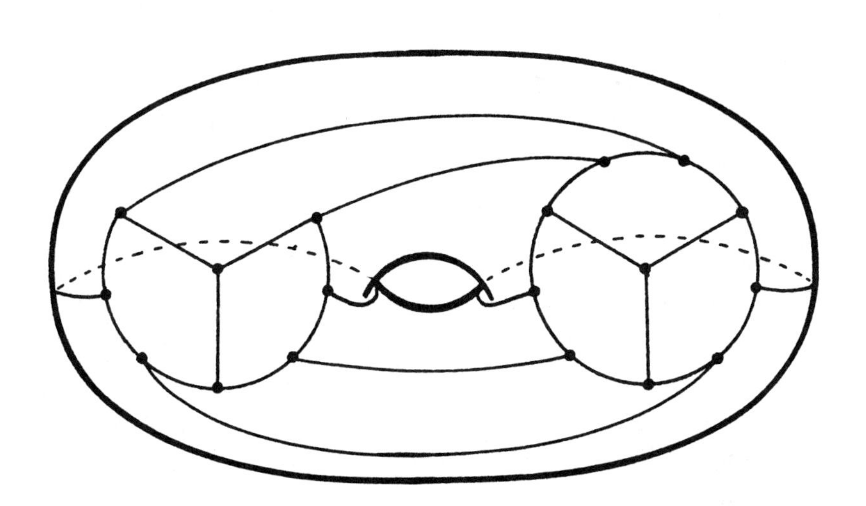

Figure 4.

2. THE DOT PRODUCT

The Blanuša snark is an example of a general construction. Two snarks F and G can be joined together to form a new snark F·G as in [2] constructed as follows:

(i) remove any two nonadjacent edges ab and cd from F;

(ii) remove any two adjacent vertices x and y from G;

(iii) join the vertices as shown in Figure 5.

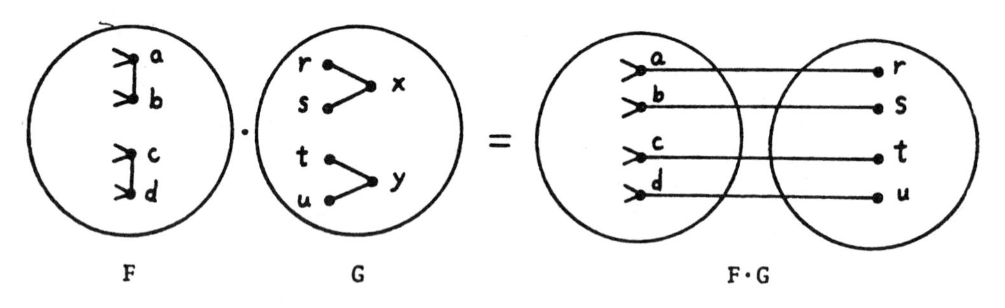

Figure 5.

Note that F·G is not uniquely determined by F and G, but depends on the particular choice of the edges and vertices removed. For example, the Blanusa snark (Figure 3) is a dot product of two Peterson graphs.

We give a bound on how fast the genus can increase under product formation. A dot product makes sense for cubic graphs in general, as long as r,s,t and u are distinct.

Theorem 1. Let F and G be cubic graphs, and let F·G be any dot product of F and G. Then their genera satisfy the inquality

$$\gamma(F \cdot G) \leqq \gamma(F) + \gamma(G) + 1 .$$

Proof. Let F and G be embedded on surfaces with genus $\gamma(F)$ and $\gamma(G)$, respectively. On the surface for F we cut out two disks containing only the edges ab and cd, adding a handle. This is now a surface of genus $\gamma(F) + 1$. On the surface for G we cut out a single disk containing only the five edges to be removed. Finally, we join these two surfaces by a tube (which does not increase the genus), sewing one end to the disk just removed and the other end to the handle by cutting out still another disk on the handle. The resulting surface has genus $\gamma(F) + 1 + \gamma(G)$ and looks like the diagram in Figure 6.

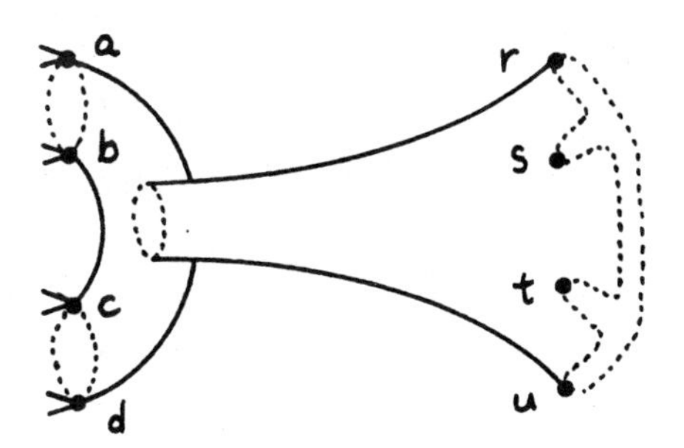

Figure 6.

It is now easy to connect a to r, and d to u, across the front of the tube, and to connect b to s, and c to t, across the back of the tube (going behind the handle). Thus, $\gamma(F \cdot G) \leqq \gamma(F) + \gamma(G) + 1$. ▯

For cubic graphs in general, this result is best possible. For example, Figure 7 shows two planar graphs with a nonplanar dot product, and the bound is attained ($\gamma = 0 + 0 + 1$).

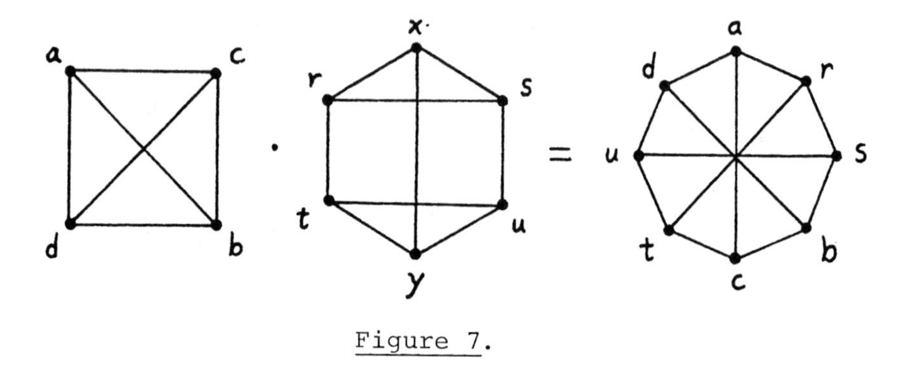

Figure 7.

However, this bound seems overly pessimistic for snarks. For example, the Blanuša snark is a dot product of two Petersen graphs, and so the theorem gives a bound of 3 when in fact the genus is 1.

More generally, we can consider those snarks which are generated by the Petersen graph under dot products. For example, Figure 8 shows a dot product of six Petersen graphs. This graph can be embedded on a surface of genus 5.

Let P^n denote any dot product of n Petersen graphs P. It is possible that $\gamma(P^n) = n - 1$, although the Szekeres snark [2] is also a dot product of six Petersen graphs, and we have as yet been unable to embed it on a surface of genus 5.

We may also consider a dot product of two Blanuša snarks (see Figure 9). We may denote such a snark by $P^2 \cdot P^2$, and observe that its genus is 2. This leads us to make the following conjecture:

$$\gamma(P^{n_1} \cdot P^{n_2} \cdots \cdot P^{n_k}) = (n_1 - 1) + \ldots + (n_k - 1)$$

$$= \sum n_i - k .$$

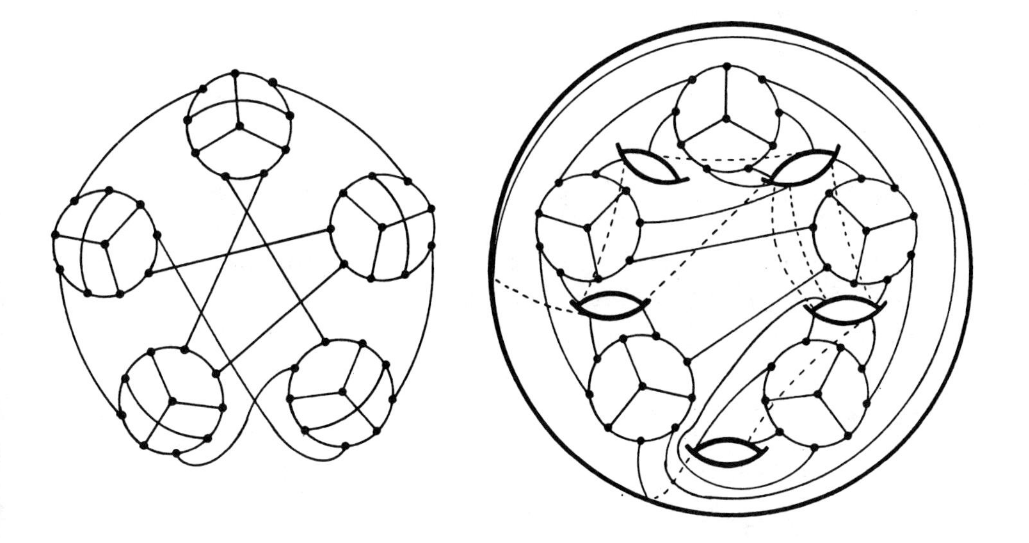

Figure 8.

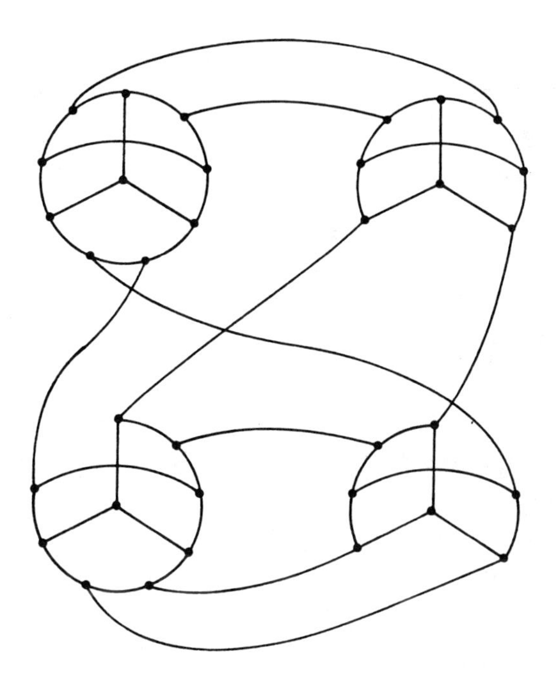

Figure 9.

3. THE FLOWER SNARKS

An infinite family of snarks was discovered independently by Isaacs and Grinberg. The <u>flower snarks</u> J_3, J_5, J_7, ... each have an odd number of 'petals' radiating from a central 'n-gon' and joined around the outside by a 'ring'. This outer ring circles the graph twice while twisting about itself much like one of Saturn's braided rings. Since it contains a triangle, J_3 is not a proper snark, but if we contract the triangle to a point, we have the Petersen snark.

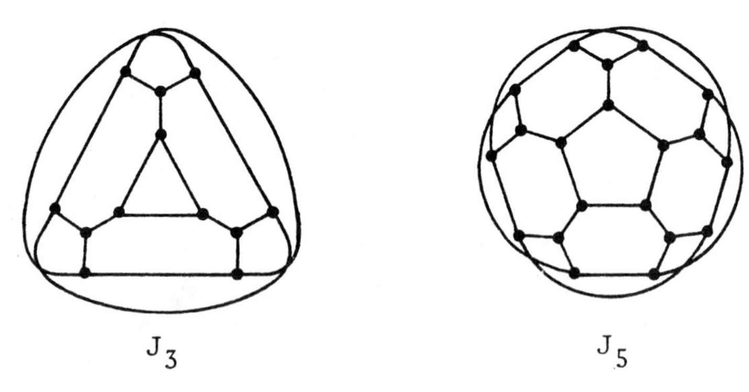

<u>Figure 10</u>.

Before obtaining the genus of the flower snarks, we shall need some terminology. We shall call the $K_{1,3}$ subgraphs that separate the petals 'triads' and name them T_1, T_2, ... , T_{2k+1} in order. Beginning at T_1, we name the edges of the central n-gon f_1, f_2, ... , f_{2k+1} in order, and the edges of the outer ring e_1, e_2, ... , e_{2k+1}, e'_1, ... , e'_{2k+1} in order (see Figure 11). The key idea that will be used in the following proof is that either e_i and e'_{i+1}, or e'_i and e_{i+1} must be transverse. The development of this idea needs some topological preliminaries.

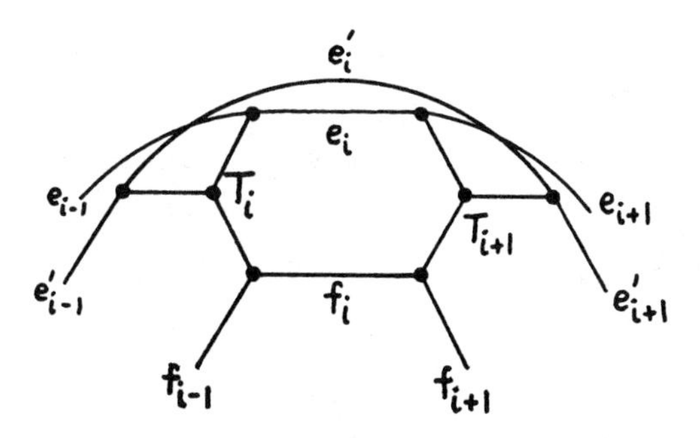

Figure 11.

Let F_m denote the sphere with m handles, and D_m the disk with m handles. Two simple closed curves c_1 and c_2 on a surface F_m are transverse if they meet at a single point and cross there. If we cut along c_1 and c_2, the classification theorem for surfaces [5, p. 75] ensures that we obtain D_{m-1}, a disk with one fewer handles. We will need the case where c_1 and c_2 meet in a segment and cross, but we still obtain D_{m-1} by cutting first along one curve and then along the other. The only difference is that the second cut is at an angle. We shall also call such curves transverse. This is illustrated in Figure 12.

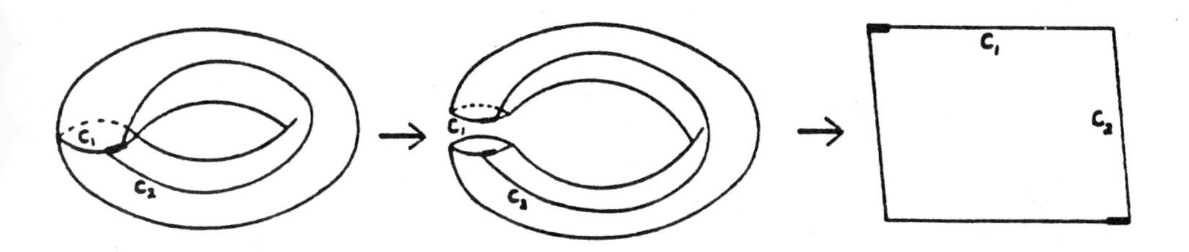

Figure 12.

We now define what it means for two curves to be transverse in D_m. An embedding of a line segment ι in D_m is called a proper embedding if the boundary and interior of ι are contained in the boundary and interior of D_m, respectively. We observe that identification of the boundary of D_m to a point yields F_m. Two properly embedded segments ι_1 and ι_2 are transverse in D_m if identification of the boundary of D_m results in two transverse simple closed curves c_1 and c_2 in F_m. We shall need the case where ι_1 and ι_2 coincide in an interval prior to intersecting the boundary of D_m, but we still obtain D_{m-1} by cutting along these segments. We shall also call such segments transverse. This is illustrated in Figure 13.

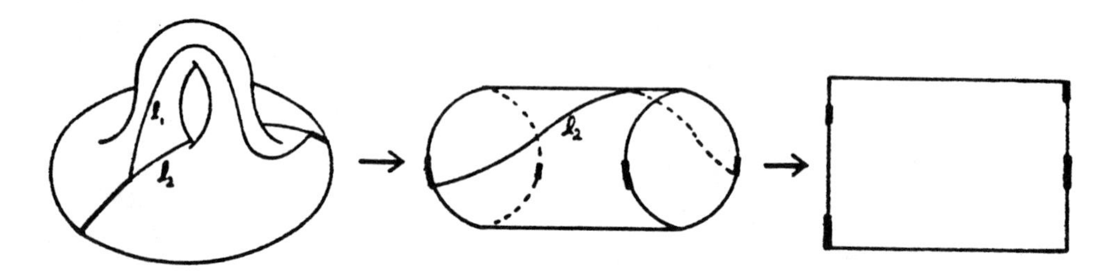

Figure 13.

We are now ready to state and prove our main result.

Theorem 2. $\gamma(J_{2k+1}) = k$, for $k = 1,2,3,\ldots$.

Proof. First we show that $\gamma(J_{2k+1}) \leq k$ by explicitly embedding J_{2k+1} on a surface of genus k. We illustrate the general process with J_7 (so that k=3). We begin by drawing J_{2k+1} on a sphere with the triad T_1 at the top, as in Figure 14, and omitting the k+1 edges

$e_1', e_3', \ldots, e_{2k+1}'$ from the outer ring. At this stage triads T_2 and T_3, T_4 and T_5, ... , and T_{2k} and T_{2k+1} are fully joined.

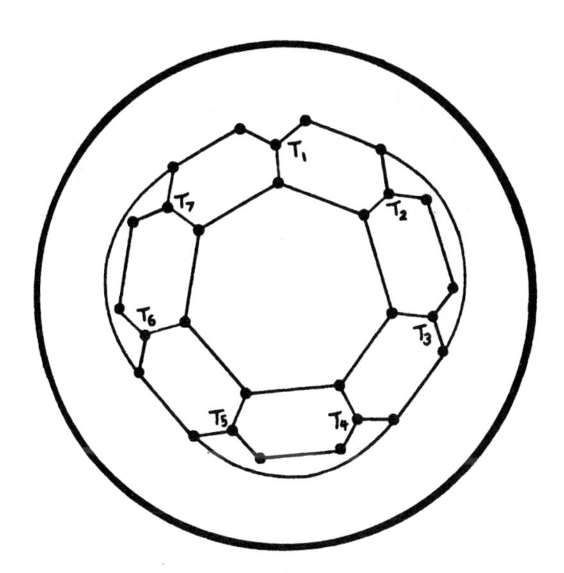

Figure 14.

Next, place a hole in each petal following an even-numbered triad, and join triads T_3 and T_4, T_5 and T_6, ... , T_{2k-1} and T_{2k}, as shown in Figure 15, by going through two holes. Finally, join triads T_1 and T_2, and triads T_1 and T_{2k+1}. Thus we have $\gamma(J_{2k+1}) \leqq k$.

We now turn to the more difficult part of the proof --showing that J_{2k+1} cannot be embedded on a surface with k-1 handles. Let F_m be the surface of minimum genus m on which J_{2k+1} can be embedded. There are three cases depending upon how the n-gon is embedded in F_m.

Case 1. The n-gon separates F_m into two components neither of which is a disk.

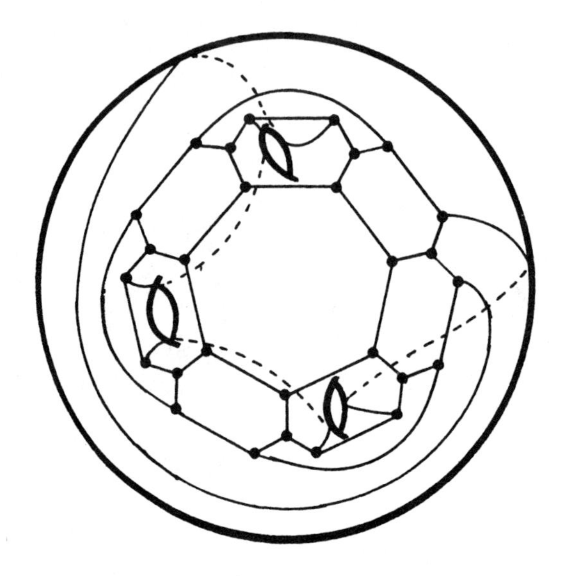

Figure 15

Since the outer ring does not intersect the n-gon, it must lie in only one of the two components. But the outer ring meets each of the triads, so all of J_{2k+1} lies in only one of the components, contradicting the minimality of m.

Case 2. The n-gon separates F_m into two components, one of which is a disk.

Considering the outer ring as in the previous case, we see that the n-gon must bound a disk. We cut out this disk along the n-gon. The result is that J_{2k+1} is embedded on a disk D_m with the n-gon along the boundary.

Consider the triads T_2 and T_3; they are joined by edges e_2 and e_2'. Without loss of generality, suppose that e_2 is adjacent to the edge of T_2 farthest from T_3 (in the sense of the oriented n-gon). Thus e_2 and e_1' are transverse in our sense. We cut along e_2 and e_1' to obtain a disk D_{m-1}. The boundary of D_{m-1} still contains edges $f_3, f_4, \ldots, f_{2k+1}$ in order, and the triads T_4, $T_5, \ldots, T_{2k+1}$ and the corresponding e_is and e_i's remain undisturbed by the cuts. In general, one of the pairs e_i and e_{i+1}', or e_{i+1} and e_i' is a transverse pair. We can cut sequentially along T_2 and T_3, T_4 and $T_5, \ldots,$ T_{2k} and T_{2k+1} eliminating a handle for each pair. Thus, we can eliminate k handles, and $m \geq k$ as desired.

Case 3. The n-gon does not separate F_m.

Since F_m is orientable, any simple closed curve in F_m is 2-sided. If all of the triads of J_{2k+1} are to the same side of the n-gon in F_m, then we cut along the other side of the n-gon and attach a disk, leaving J_{2k+1} embedded in F_{m-1}, contradicting the minimality of m. Thus, we may assume that there are triads on both sides of the n-gon. We may further assume that T_1 and T_2 are on opposite sides of the n-gon in F_m. Then the n-gon and e_1 are transverse in our sense. We cut along these curves to obtain a disk D_{m-1}. The boundary D_{m-1} consists of two copies each of the n-gon and e_1. We illustrate this in Figure 16.

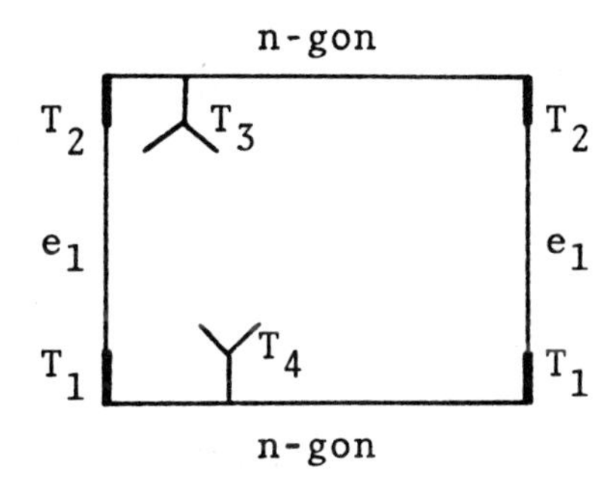

Figure 16.

Triads on different sides of the n-gon in F_m attach to different copies of the n-gon in the boundary of D_{m-1}, as in Figure 16.

Consider T_4 and T_5. As before, we claim that either e_3 and e_4', or e_4 and e_3' are transverse in our sense. If T_3, T_4 and T_5 are attached to the same copy of the n-gon, then this reduces to the previous case. If not, the result is still true; we illustrate one of the three possible patterns in Figure 17.

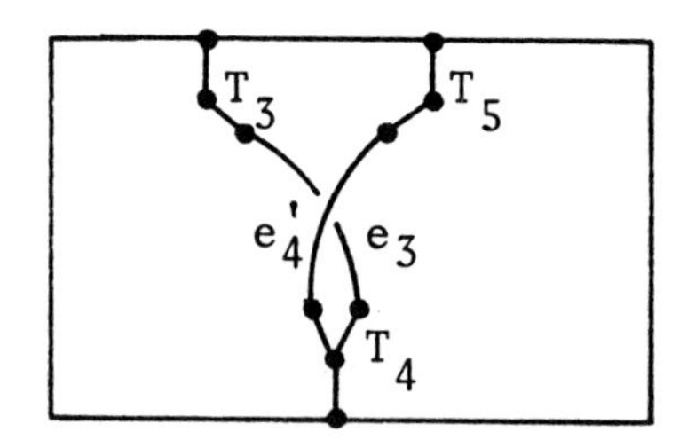

Figure 17.

As before, take the edge of T_4 closer to T_3. Without loss of generality, it is joined to T_5 by e_4', and so e_3 and e_4' are transverse. Cutting along these curves reduces D_{m-1} to D_{m-2} and leaves the boundary of D_{m-1} undisturbed along both copies of the n-gon from f_5 to f_{2k+1}. Also undisturbed are the triads T_6 to T_{2k+1}. Thus, we can eliminate one handle for each of the pairs T_4 and T_5, T_6 and T_7, ... , T_{2k} and T_{2k+1}. This eliminates k-1 handles, so $m - 1 \geq k - 1$ and it follows that $m \geq k$. ▯

4. THE GOLDBERG SNARKS

We now give embeddings of the infinite family G_3, G_5, G_7, ... known as the Goldberg snarks [2]. Like the smallest of the flower snarks, G_3 is not a proper snark since it contains a triangle, but becomes one if the triangle is contracted to a point. We draw G_3 and G_5 in Figure 18.

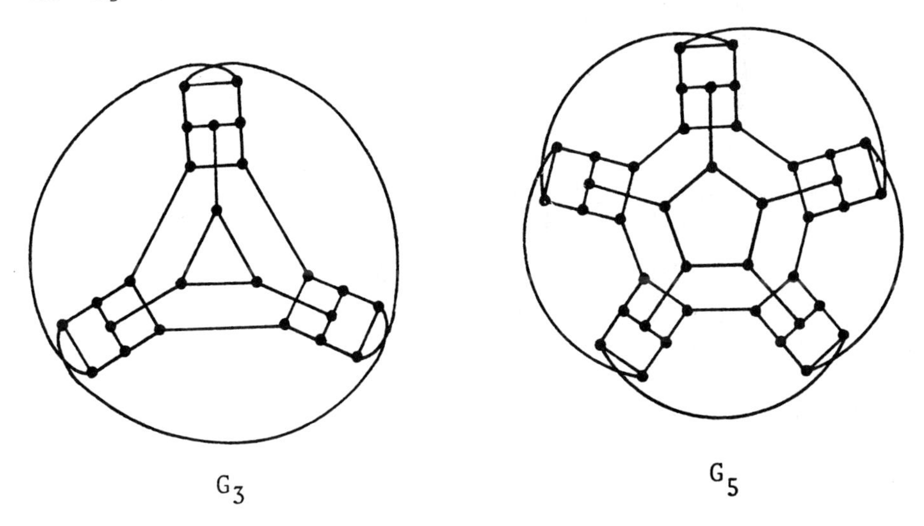

Figure 18.

The following theorem gives an upper bound on the genus of G_n. We suspect that equality holds.

Theorem 3. $\gamma(G_n) \leqq n-1$ for $n = 3,5,7,\ldots$.

Proof. We embed G_n on a surface of genus n-1, and illustrate the general embedding with G_5 in Figure 19. □

Figure 19.

REFERENCES

1. D. Blanusa, Problem ceteriju boja (The problem of four colors), Hrvatsko Prirodoslovno Drustvo Glasnik Mat.-Fiz. Astr. Ser. II, 1(1946) 31-42.
2. A. G. Chetwynd and R. J. Wilson, Snarks and super snarks, The Theory and Applications of Graphs (G. Chartrand, ed.) Wiley, New York (1981) 215-241.
3. M. Gardner, Mathematical games, Scientific American 234(1976) 126-130.
4. A. B. Kempe, A Memoir on the theory of mathematical form, Phil. Trans. Roy. Soc. London 177 (1886) 1-70.
5. J. Stillwell, Classical Topology and Combinatorial Group Theory, Springer-Verlag, New York (1980).
6. P. G. Tait, Remarks on the colouring of maps, Proc. Roy. Soc., Edinburgh 10 (1880) 729.

USING GRAPHS TO INVESTIGATE THE AUTOMORPHISM GROUPS OF NILPOTENT GROUPS

Ursula Martin Webb

Department of Mathematics
University of Illinois
Urbana, Illinois 61801

ABSTRACT. In this paper we survey results about automorphisms of finite and infinite nilpotent groups that have been obtained using graphical techniques. We prove that if H is any finite group then there is a finitely generated torsion-free nilpotent group G whose automorphism group is an extension of a torsion-free nilpotent group of class 2 by H.

1. INTRODUCTION

The object of this paper is to describe some of the ways in which graphs have been used to study automorphisms of nilpotent groups. The underlying idea is that whereas it is often quite difficult to construct a group with given properties, the corresponding problem for graphs may be much easier. So to find a group with given automorphism group, one first finds a graph with that automorphism group, then associates a nilpotent group to the graph, in such a way that the automorphism group of the group is related to that of the graph. Similar ideas have been used in other areas of algebra and combinatorics; see Babai [1] for a detailed account.

We give in section 3 an account of results about automorphisms of finite and torsion-free nilpotent groups which have been obtained by graphical methods. We shall illustrate these methods in section 4 by proving the following theorem.

Theorem A: Let H be any finite group. Then there is a finitely generated torsion-free nilpotent group G of nilpotency class 3 such that Aut G is an extension of a torsion-free nilpotent group of class 2 by H.

Theorem A is a consequence of the more precise result Theorem B, which is stated in section 3.

By applying Theorem A in the case H = 1 we obtain the following corollary.

Corollary: There exists a finitely generated nilpotent group whose automorphism group is finitely generated, torsion-free and nilpotent of class 2.

This result provides the first known examples of groups with such automorphism groups.

I should like to thank Rebekka Struik and Derek Robinson for their helpful comments on this work.

2. SOME BACKGROUND INFORMATION ON NILPOTENT GROUPS

We describe the groups in which we are interested, and what is known about their automorphism groups.

If x and y are elements of the group G then we denote their commutator $x^{-1}y^{-1}xy$ by $[x,y]$. If H and K are subgroups of G, [H,K] denotes the subgroup generated by $\{[h,k] \mid h \in H, k \in K\}$. The lower central series of G is the descending chain of normal subgroups defined by $\gamma_1(G) = G$, $\gamma_2(G) = [G,G]$ and $\gamma_{i+1}(G) = [\gamma_i(G),G]$ for each i. If $\gamma_{n+1}(G) = 1$ and $\gamma_n(G) \neq 1$, G is said to be nilpotent of class n. A nilpotent group of class n is called an $N_n(0)$ group if each non-trivial factor $\gamma_i(G)/\gamma_{i+1}(G)$ is free abelian, and an $N_n(p)$ group, where p is a prime, if each non-trivial $\gamma_i(G)/\gamma_{i+1}(G)$ is an elementary abelian p-group. In either case we let $\bar{G} = G/\gamma_2(G)$, and denote the minimum number of generators of $\bar{G}$ by d(G).

An element of G of the form $[..[[y_1,y_2]y_3]...]y_n]$ is called a left normed commutator of weight n, and written as $[y_1,...,y_n]$. If $X = \{x_k | k \in D\}$ is a set of generators for G then $\gamma_i(G)$ is generated modulo $\gamma_{i+1}(G)$ by all left normed commutators of weight i with entries drawn from X. We shall be interested in groups which are nilpotent of class 3. In such a group $\gamma_2(G)$ is abelian (see Hall [3], Lemma 3.7 for example) and the following identities, which are easily proved from the usual commutator identities (2.1 of [3]), hold for all a, b, c and d in G and r, s, t in $\mathbb{Z}$.

$$[ab,c,d] = [a,c,d][b,c,d]$$
$$[a,bc,d] = [a,b,d][a,c,d]$$
$$[a,b,cd] = [a,b,c][a,b,d]$$
$$[a^r,b^s,c^t] = [a,b,c]^{rst}$$
$$[a,b,c] = [b,a,c]^{-1}$$
$$[a,b,c][c,a,b][b,c,a] = 1 \quad \text{(the Witt identity)}$$

We let Z(G) denote the center of G, that is $\{g \in G | xg=gx$ for all $x \in G\}$.

We now consider the automorphism group of a group G. Any automorphism of G preserves $\gamma_2(G)$ so induces a homomorphism

$$\alpha : \text{Aut } G \to \text{Aut}(\bar{G}) .$$

We denote the image of α by $A(G)$, and its kernel by $K(G)$, so that the sequence

$$1 \to K(G) \to \text{Aut } G \to A(G) \to 1$$

is exact. Notice that any element of G induces an automorphism of G by conjugation: the group of such automorphisms forms a subgroup of $K(G)$ denoted by Inn G, which is isomorphic to $G/Z(G)$. If G is an $N_n(p)$ group it follows from the corollary on page 17 of [3] and an easy induction argument that $K(G)$ is an $N_{n-1}(p)$ group, and $A(G)$ is isomorphic to a subgroup of $\text{Aut}(\bar{G}) = GL(d(G),p)$. If G is an $N_n(0)$ group, it follows in a similar way that $K(G)$ is an $N_{n-1}(0)$ group, and $A(G)$ is isomorphic to a subgroup of $GL(d(G),\mathbb{Z})$. Thus to determine Aut G completely we need to know what groups can occur as $A(G)$, and this is what is described in the next section.

3. AUTOMORPHISMS OF NILPOTENT GROUPS

In this section we describe what is known about $A(G)$, and the related question of what abelian groups and torsion-free nilpotent groups can occur as automorphism groups.

Graphical methods were first used by H. Heineken and H. Liebeck [4] to study $N_2(p)$ groups. They showed that if p is an odd prime and H is a finite p-group then there is a finitely generated $N_2(p)$ group G with $A(G) \cong H$. This result was extended by the author to the case $p = 2$, and to H finitely generated [10]. (The case $p = 2$ was also considered by Hughes [5] and Zurek [13].) In fact the method allows us to construct many groups with the required property, since it was shown in [10] that if Γ is any graph on n vertices which satisfies a rather weak technical condition then there is an $N_2(p)$ group G with $d(G) = n$ and $A(G) \cong \text{Aut } \Gamma$.

We mention in particular the case when Γ is a graphical regular representation or GRR. A graph Γ with automorphism group isomorphic to C is called a GRR of C if $\text{Aut } \Gamma$ acts freely (that is, no non-identity element has any fixed points) and transitively on the vertices of Γ. (In other words, the vertices afford the regular representation of $C \cong \text{Aut } \Gamma$.) The classification of finite groups with a GRR has been completed recently by Hetzel, Godsil and others (see [2]): it turns out that any finite group has a GRR unless it is abelian of exponent at least 3, generalized dicyclic or one of 13 small exceptions. It was shown in [10], Theorem 5.2, that if C is a finite group with a GRR then there is an $N_2(p)$ group G with $d(G) = |C|$ and $A(G) \cong C$, in which $\bar{G}$ affords the regular representation of C.

Similar techniques have been applied to $N_2(0)$ groups. However we cannot expect such strong results in this case, since it is not hard to check that if G is an $N_2(0)$ group then $A(G)$ contains an element of order 2, namely the extension to G of the automorphism of the desired quotient which inverts every element. So for such groups the following result, obtained in U. Webb [11], is the best to be hoped for. (The case H finite was considered by H. Liebeck in [8].)

Theorem. Let R be any group. Then there is an $N_2(0)$ group G with $A(G)$ isomorphic to an elementary abelian

2-group of rank $d(G)$ extended by R. If R is finite, G can be chosen to be finitely generated.

To realize any finite group as $A(G)$, for G a torsion-free nilpotent group, we need to turn to groups of class 3, and this is the point of our main theorem.

Theorem B. Let $\mathcal{J}$ be any finite group with a GRR, Γ. Let H be any subgroup of $\mathcal{J}$. Then there is an $N_3(0)$ group $G = G(\Gamma)$ with $d(G) = |\mathcal{J}|$ satisfying

1) $A(G) \cong H$
2) $A(G/\gamma_3(G))$ is isomorphic to an extension of an elementary abelian two-group of rank $d(G)$ by $\mathcal{J}$.

Observe that since any finite group can be realized as a subgroup of a group with a GRR (for example a suitable symmetric group), Theorem A is an immediate consequence of Theorem B.

In fact it is clear from the proof of Theorem B, and of Lemma 1 which precedes it and describes the automorphism group of an $N_3(0)$ group constructed from an arbitrary graph, that many variations of the above result are true. For example the result may be extended to other finite and infinite groups $\mathcal{J}$. We can also describe $K(G)$ quite precisely in terms of Γ. In particular by choosing Γ carefully it can be shown that $K(G) \cong \mathrm{Inn}\ G \cdot C(G)$ where $C(G)$ is the group of automorphisms of G which act trivially on $G/\gamma_3(G)$. It is well known that $C(G)$ is isomorphic to $\mathrm{Hom}(\bar{G}, \gamma_3(G))$.

We now consider groups with abelian automorphism group. Groups with elementary abelian automorphism group are rather numerous, for if G is an $N_2(p)$ group with $A(G) = 1$ then $\mathrm{Aut}\ G = K(G)$, which is an elementary abelian p-group. It follows from the results described above, and the fact that almost all graphs have trivial automorphism group and satisfy the weak technical condition mentioned, that almost all p-groups constructed by this method have elementary abelian automorphism group. In fact rather more is true, for it has recently been proved (Webb [12]) that as $n \to \infty$ the proportion of $N_2(p)$ groups G with $d(G) = n$ having $A(G) = 1$ tends to 1. However it is rather hard to construct groups with $A(G) = 1$ by purely group theoretic methods: see for instance Iyer [6], Struik [9] and Jonah and Konvisser [7].

The picture is very different when we consider torsion-free groups which might occur as automorphism groups, for it is not hard to see that no group can have free abelian automorphism group. For such a group G would have $G/Z(G) = \text{Inn } G \leq \text{Aut } G$, so that $G/Z(G)$ would be free abelian. But then G would admit an automorphism of order two extending the automorphism of $G/Z(G)$ which inverts each element, so that Aut G would not be torsion free. Thus one is led to ask whether a group can have a torsion-free nilpotent automorphism group; we see by applying Theorem B in the case $H = 1$ that the answer is yes. We remark for completeness that a group can have a finitely generated infinite abelian automorphism group with elements of finite order; in particular if p is any odd prime the group $\mathbb{Z}^{20} \times (\mathbb{Z}/p\mathbb{Z})^{36}$ can occur as an automorphism group (Webb, unpublished).

4. PROOF OF THEOREM B.

The theorem will follow from the following construction. We shall describe how to associate to a graph Γ an $N_3(0)$

group $G(\Gamma) = G$. Lemma 1 describes Aut G and by considering the case when Γ is a GRR we obtain Theorem B.

We suppose that Γ is a finite or infinite connected graph on vertices V which are totally ordered by $<$, with edges E. Suppose we are given a map $\gamma : V \to \mathbb{N}$, the natural numbers, and an injection $\delta : V \to E$ which assigns to each vertex v a distinct edge $v\delta = \{v\delta_1, v\delta_2\}$ with $v\delta_1 < v\delta_2$ such that the following conditions hold.

(1) Either

a) Γ is not complete, each vertex has finite degree $d \geq 2$ and Aut Γ acts freely on V (that is each non-identity element has no fixed points)

or

b) Each vertex of Γ has degree at least 2 and Γ has no circuit of length 3 or 4. (Notice that any GRR satisfies 1(a).)

(2) For each vertex v in V, $v\delta$ lies in a circuit of Γ with no vertex adjacent to v.

(3) The maps δ_1 and δ_2 commute with Aut Γ, that is $(v\delta_1)\pi = (v\pi)\delta_1$, $(v\delta_2)\pi = (v\pi)\delta_2$ for each $v \in V$, $\pi \in$ Aut Γ. Notice that this means that Aut Γ preserves $E_1 = V\delta$ and $E_2 = E \backslash E_1$.

We now define G as the group on generators $\{x_v \mid v \in V\}$ with relators $X \cup Y \cup Z$ given as follows:

$$X = \{[x_i, x_j] \mid \{i,j\} \in E_2\}$$

$$Y = \{[x_i, x_j][x_k, x_i, x_j]^{k\gamma} \mid k \in V,\ i = k\delta_1,\ j = k\delta_2\}$$

$$Z = \{[x_r, x_s, x_u, x_t] \mid r,s,t,u \in V\} .$$

Observe first that the relators Z ensure that $_{4}(G) = 1$, so G is nilpotent of class at most 3. Furthermore $G/\gamma_3(G)$ is just the group defined on page 404 of [11], and is independent of γ and δ. Thus we know immediately that $\bar{G}$ is free abelian and has a basis $\{\bar{x}_i | i \in V\}$, and that $\gamma_2(G)/\gamma_3(G)$ is free abelian with a basis $\{[x_i,x_j]\gamma_3(G)\ \{i,j\} \notin E,\ i < j\}$.

It follows from section 2 that $\gamma_2(G)$ is abelian, and generated by all left-normed commutators of weight 2 or 3. Careful examination of the relators and use of the commutator identities of section 2 shows that the only dependence relations between these generators are consequences of the following

$$[x_i,x_j] = 1 \qquad \text{for } i < j,\ \{i,j\} \in E_2$$

$$\left.\begin{array}{r} [x_i,x_j][x_k,x_i,x_j]^{k\gamma} = 1 \\ [x_j,x_i][x_k,x_j,x_i]^{-k\gamma} = 1 \end{array}\right\} \qquad \text{for } i = k\delta_1,\ j = k\delta_2$$

$$[x_i,x_j,x_k] = 1 \qquad \text{if } \{i,j,k\} \text{ involves at least 2 edges}$$

$$[x_i,x_j,x_k][x_k,x_i,x_j][x_j,x_k,x_i] = 1 \qquad \text{if } \{i,j,k\} \text{ involves no edges}$$

$$\left.\begin{array}{r} [x_i,x_j,x_k] = 1 \\ [x_k,x_i,x_j][x_j,x_k,x_i] = 1 \end{array}\right\} \qquad \text{if } \{i,j\} \text{ is an edge and } \{i,k\} \text{ and } \{j,k\} \text{ are not}$$

It is clear from this that $\gamma_2(G)$ is free abelian on generators $\{[x_i,x_j] | i < j,\ \{i,j\} \notin E\}$ together with certain left normed commutators of weight 3. Hence G is an $N_3(0)$ group. (It follows from property (1) of Γ that $\gamma_3(G) \neq 1$.)

To determine Aut G we observe that any automorphism

of G induces an automorphism of $G/\gamma_3(G)$, so that A(G) embeds in $A(G/\gamma_3(G))$. Since Γ satisfies (1), we know from Corollary (3) of [11] that $A(G/\gamma_3(G))$ is isomorphic to $D \rtimes \mathrm{Aut}\,\Gamma$, where D is the elementary abelian 2-group consisting of automorphisms of $G/\gamma_3(G)$ which act on $G/\gamma_2(G)$ by inverting some of the generators $\bar{x}_i$ and leaving the rest fixed. Hence A(G) is isomorphic to a subgroup of $D \rtimes \mathrm{Aut}\,\Gamma$. The function of δ and γ is to determine this subgroup, as the next lemma shows.

Lemma 1. Let Γ and G be as above. Let

$$A_\gamma = \{\beta \in \mathrm{Aut}\,\Gamma \mid (v\beta)\gamma = v\gamma \quad \text{for all} \quad v \in V\} .$$

Then $A(G) \simeq A\gamma$.

<u>Proof</u>. Let α be an automorphism of G. Since A(G) embeds in $A(G/\gamma_3(G))$ we may assume that for each i

$$x_i\alpha = x_{i\pi}^{\varepsilon_{i\pi}} u_{i\pi}$$

where $\varepsilon_{i\pi} = \pm 1$, $u_{i\pi} \in \gamma_2(G)$ and π is an automorphism of Γ. For notational convenience let $\phi = \pi^{-1}$, so that

$$x_{i\phi}\alpha = x_i^{\varepsilon_i} u_i .$$

It follows from our remarks above that u_i can be uniquely expressed as

$$u_i = \prod [x_r, x_s]^{{}_i\alpha_{rs}} w_i$$

where $w_i \in \gamma_3(G)$, ${}_i\alpha_{rs} \in \mathbb{Z}$ and the product is over all pairs $\{r,s\}$ with $\{r,s\}$ not an edge, with $r = i$ if $i \in \{r,s\}$ and $r < s$ otherwise.

Since ϕ preserves E_2 it follows that if $\{i,j\} \in E_2$ then

$$[x_{i\phi}, x_{j\phi}] = 1 = [x_i, x_j].$$

Then

$$[x_{i\phi}\alpha, x_{j\phi}\alpha] = 1$$

and hence

$$u = 1 \qquad (\dagger)$$

where

$$u = [x_i^{\varepsilon_i}, u_j][u_i, x_j^{\varepsilon_j}]$$

$$= [x_i^{\varepsilon_i}, \prod [x_r,x_s]^{{}_j\alpha_{rs}}][\prod [x_r,x_s]^{{}_i\alpha_{rs}}, x_j^{\varepsilon_j}].$$

Similarly if $k \in V$ and $k\delta_1 = i$, $k\delta_2 = j$ then $k\delta_1\phi = i\phi = (k\phi)\delta_1$, $k\delta_2\phi = j\phi = (k\phi)\delta_2$ and we have

$$[x_{i\phi},x_{j\phi}][x_{k\phi},x_{i\phi},x_{j\phi}]^{(k\phi)\gamma} = 1$$

and applying α, and using the relation

$$[x_i,x_j][x_k,x_i,x_j]^{k\gamma} = 1$$

we obtain

$$[x_k\, x_i\, x_j]^{-k\gamma\, \varepsilon_1\, \varepsilon_j}\, u [x_k\, x_i\, x_j]^{(k\phi)\gamma\varepsilon_i\varepsilon_j\varepsilon_k} = 1 \qquad (\dagger\dagger)$$

We now suppose that t is any vertex of Γ, and that $\{i,j\}$ is any edge such that $\{i,t\}$ and $\{j,t\}$ are not edges. Then $[x_i,x_t,x_j] \neq 1$, and $[x_k,x_t,x_j][x_t,x_j,x_i]=1$. Equating coefficients of $[x_i,x_t,x_j]$ in $(\dagger)$ and $(\dagger\dagger)$ we find that

$$\varepsilon_j\, {}_i\alpha_{it} = \varepsilon_i\, {}_j\alpha_{jt} \quad \text{if} \quad \{i,j\} \neq \{t\delta_1,t\delta_2\}$$

$$t\gamma\, \varepsilon_i\varepsilon_j - \varepsilon_i\, {}_j\alpha_{jt} + \varepsilon_j\, {}_i\alpha_{it} = (t\phi)\gamma\, \varepsilon_i\varepsilon_j\varepsilon_t$$

if

$$i = t\delta_1, \quad j = t\delta_2.$$

By assumption (2), $\{t\delta_1, t\delta_2\}$ lies in a circuit consisting of vertices which are not adjacent to t. By applying the first equation repeatedly tc the edges of this circuit we find that when $i = t\delta_1$, $j = t\delta_2$

$$\varepsilon_i \, {}_j\alpha_{jt} = \varepsilon_j \, {}_i\alpha_{it}$$

so that from the second equation

$$t\gamma = (t\phi)\gamma \, \varepsilon_t.$$

Since $t\gamma$ and $t\phi\gamma$ are both positive, $\varepsilon_t = 1$ so $t\phi\gamma = t\gamma$. Since this holds for any choice of t, $\pi \in A_\gamma$, and the map $\alpha \to \pi$ induces an embedding of $A(G)$ into A_γ.

Conversely suppose that $\pi \in A_\gamma$. It is easy to check that the map $\alpha : G \to G$ by $a : x_i \to x_{i\pi}$ for each i is injective and surjective, and preserves the given relations, so α is an automorphism of G. Hence $A(G) \simeq A_\gamma$. Thus the lemma is proved. □

We note that the method of the lemma allows us to determine the coefficients ${}_i\alpha_{rs}$, and hence to describe K(G) exactly in terms of Γ. We shall not do this, but remark that if Γ has the property that for each vertex t the graph obtained by deleting t and the vertices adjacent to it is connected, then $K(G) = \text{Inn } G \cdot C(G)$, where C(G) consists of those automorphisms which act as the identity of $F/\gamma_3(G)$.

Now we prove Theorem B.

<u>Proof of Theorem B</u>. We let Γ be any GRR of $\mathcal{J}$, so that Γ satisfies 1(a). We need only to describe how to construct δ and γ, so that δ satisfies (2) and (3) and $A_\gamma = H$. Since Γ is a GRR we may assume that the vertices of Γ are labeled by the elements of $\mathcal{J}$ and if

$\pi \in \mathcal{J}$ the automorphism π^* of Γ induced by π acts by $g\pi^* = \pi^{-1}g$ for each $g \in \mathcal{J}$. Let the degree of Γ be d.

We explain first the construction of δ. Since Γ is a GRR each vertex has degree d at least 3 and at most $k-4$, where $k = |\mathcal{J}|$. Let N be the graph on the d vertices adjacent to 1 (the vertex representing the identity of $\mathcal{J}$) in Γ, and M the graph on the $k-(d+1)$ vertices which are not 1 or vertices of N. Suppose that M and N have m and n edges respectively, and that f edges have one end in M and the other in N. I claim that either M or the complement of N contains a circuit. For if not then M and the complement of N are both forests (a forest is a graph with no circuits), so that

$$m \leq k - (d+1) - 1$$

and

$$n \geq \binom{d}{2} - (d-1).$$

Now the number of ends of edges in N is

$$d + 2n + f = d^2$$

so that

$$f + n = d^2 - d - n \leq \binom{d}{2} + d - 1.$$

The total number of edges of Γ is $dk/2$, so

$$\begin{aligned} dk/2 &= d + f + n + m \\ &\leq d + \binom{d}{2} + d - 1 + k - d - 2 \end{aligned}$$

so that

$$k(d-2) \leq (d-2)(d+3)$$

which is impossible as

$$3 \leq d \leq k - 4.$$

Now suppose that M contains a circuit, S. Pick an edge $\{h,k\}$ in this circuit, and define $1\delta_1 = h$, $1\delta_2 = k$.

Now for each $g \in G$, define $g\delta_1 = gh$, $g\delta_2 = gk$. Then the automorphism $(g^{-1})^*$ of Γ sends 1 to g, and S to a circuit with no vertex adjacent to g, and containing the edge $\{gh, gk\}$. Thus (2) is satisfied. For (3) observe that $(g\delta_1)\pi^* = (\pi^{-1}g)\delta_1 = \pi^{-1}gh = (g\pi^*)\delta_1$.

Suppose M does not contain a circuit, but that the complement of N does. Replace Γ by its complement, which is still a GRR of $\mathcal{J}$ satisfying 1(a). In this new graph the circuit in the complement of N becomes a circuit on the vertices not adjacent to 1, so that we may define δ as above.

Now we define γ. Let $H \leq \mathcal{J}$, and let $Hu_1, \ldots, Hu_n$ be the distinct right cosets of H in $\mathcal{J}$. Define $\gamma : \Gamma \to \mathbb{N}$ by $g\gamma = k$ if $g \in Hu_k$. Then $(g\pi^*)\gamma = g\gamma$ for all g if and only if $\pi \in H$, so $A_\gamma = H$ and the theorem is proved. □

REFERENCES

1. L. Babai, On the abstract group of automorphisms, Combinatorics, ed. by H.N.V. Temperley CUP, London, (1981), 1-40.
2. C. Godsil, GRR's for non-solvable groups, Colloquia Mathematica Societatis János Bolyai, 25, North-Holland, Amsterdam, (1981), 221-239.
3. P. Hall, The Edmonton notes on nilpotent groups, Queen Mary College Math, Notes, London, 1969 MR 44#316.
4. H. Heineken and H. Liebeck, The occurrence of finite groups in the automorphism group of nilpotent groups of class 2, Arch. Math. 25 (1974), 8-16. MR50#2 337,20E15.

5. A. Hughes, Automorphisms of nilpotent groups and super soluble groups, The Santa Cruz conference on finite groups, American Mathematical Society, Providence, (1980), 205-208.
6. H. Iyer, Finite groups with elementary abelian automorphism groups, manuscript.
7. D.W. Johan and M.W. Konvisser, Some non-abelian p-groups with abelian automorphism groups, Arch. Math. 26(1975), 131-133. MR51#3301 20D45.
8. H. Liebeck, Automorphism groups of torsion-free nilpotent groups of class two, J. London Math. Soc. (2) 10 (1975), 349-356. MR51#13063 20F55.
9. R.R. Struik, Some non-abelian 2-groups with abelian automorphism groups, Arch. Math. 39 (1982), 299-302.
10. U. Webb, The occurrence of groups as automorphisms of nilpotent p-groups, Arch. Math. 37 (1981), 481-498.
11. U. Webb, An independence theorem for automorphisms of torsion-free groups, Math. Proc. Cambridge Philos. Soc. 90 (1981), 403-409. MR82k#20057.
12. U. Webb, On the automorphism group of a variety, to appear.
13. G. Zurek, Eine Bemerkung zu einer Arbeit von Heineken und Liebeck, Arch. Math. 38 (1982), 206-207.